ADVANCED LEVEL

Practical Work for

CHEMISTRY

ADVANCED LEVEL

Practical Work for

CHEMISTRY

Andrew Hunt

Hodder & Stoughton

A MEMBER OF THE HODDER HEADLINE GROUP

Acknowledgements

The Publishers would like to thank the following for permission to reproduce copyright material:
Fig 1.1 Lester Lefkowitz/CORBIS; Fig 12.1 taken from *Introducing Measurement Uncertainty*, Vicki Barwick &
Elizabeth Prichard, LGC, 2003.

Every effort has been made to trace all copyright holders, but if any have been inadvertently overlooked the Publishers
will be pleased to make the necessary arrangements at the first opportunity.

Although every effort has been made to ensure that website addresses are correct at time of going to press, Hodder
Murray cannot be held responsible for the content of any website mentioned in this book. It is sometimes possible to
find a relocated web page by typing in the address of the home page for a website in the URL window of your browser.

Papers used in this book are natural, renewable and recyclable products. They are made from wood grown in
sustainable forests. The logging and manufacturing processes conform to the environmental regulations of the country
of origin.

Orders: please contact Bookpoint Ltd, 130 Milton Park, Abingdon, Oxon OX14 4SB. Telephone: (44) 01235 827720. Fax:
(44) 01235 400454. Lines are open from 9.00–6.00, Monday to Saturday, with a 24 hour message answering service. Visit
our website at www.hodderheadline.co.uk.

© Andrew Hunt 2004
First published in 2004 by
Hodder Murray, a member of the Hodder Headline Group
338 Euston Road
London NW1 3BH

Impression number   10 9 8 7 6 5 4 3 2 1
Year   2010 2009 2008 2007 2006 2005 2004

Cover photo from Science Photo Library
Typeset in 10/12pt ITC Century Book by Tech-Set Ltd, Gateshead, Tyne & Wear
Printed in Great Britain by Martins, The Printers, Berwick upon Tweed

A catalogue record for this title is available from the British Library

ISBN 0 340 88672 2

CONTENTS

PREFACE

This book will help you to succeed with practical chemistry. The chapters cover all the main topics in advanced level courses. The guidance will be equally helpful whether your work is being tested by practical exams, written papers or teacher assessment.

In all your practical work you are expected to develop your skills in:

- planning
- implementing
- analysing and concluding
- evaluating.

Each chapter in this book shows what is expected of you under these headings in each of the main areas of your laboratory work.

The book covers the requirements of all the Awarding Bodies. You may not be expected to tackle everything in the book so you should check carefully with the specification for your course.

None of the information in this book is intended as instructions for practical work so there are no detailed lists of safety precautions. Always carry out a risk assessment when planning your own experiments and investigations. All your plans should then be checked by a qualified teacher or supervisor.

The advice in this book draws on the experience of all the teachers and examiners who have inspired a varied programme of laboratory activities for chemistry at this level. I have drawn on many sources especially those listed in Chapter 13.

I am grateful to detailed advice from Ian Brandon and Peter Borrows during the planning of this book. I would also like to thank Katie Blainey, Becca Law and the team at Hodder Murray for their expert input.

Andrew Hunt

2004

CHAPTER ONE

Practical skills

Introduction

Chemistry is an experimental subject. Theory does not make sense in the abstract but it does when it is seen as an attempt to explain observations in the laboratory and in the natural world.

The skills of laboratory chemistry are of great practical importance too. Chemistry is about making useful new chemicals (synthesis) and finding out what materials are made of (analysis).

The practical work in your advanced chemistry course will help you to:

- use chemical knowledge and understanding to explain the phenomena you observe
- make predictions based on theory and then test your predictions experimentally
- develop the practical skills needed to make relevant observations, to carry out analysis accurately and to prepare new chemicals cleanly, safely and with good yields
- understand the ways of planning, carrying out and interpreting practical investigations.

Figure 1.1 You will learn to carry out standard procedures safely with precision and accuracy

Your practical knowledge and skills will be assessed either by external examinations, by your teacher, or by a mixture of these two approaches. You should study the specification for your course and, if relevant, past exam papers with the mark schemes. Consult your teacher to find out how your own work will be judged.

The Awarding Bodies write their course specifications for teachers and examiners. Sometimes the meaning can be obscure to students during a chemistry course. This book will help you appreciate what you have to do to get high marks.

This chapter provides an overview of the standards for experimental and investigative work. Later chapters show in more detail how these standards apply to specific areas of chemistry.

Whichever course you are following, you will be assessed on your performance in four skill areas:

- planning
- implementing
- analysing evidence and drawing conclusions
- evaluating evidence and procedures.

Advanced courses are examined at two levels, AS and A2. The same skills are tested at both levels. In your AS course the practical assessment is based on the content of just the AS modules. A2 practical work for assessment is related to the main A2 theory modules but you may well need to draw on AS experience as well.

1.2 Planning

At first planning may seem very difficult because it can be hard to know what is possible when you have limited experience of chemistry. You need to bear in mind that your teachers and the examiners know from the course specification what you have done at each stage of your work and what it is reasonable for you to suggest in your plans.

The crucial point is that when planning any practical activity you should expect to use techniques, chemicals and equipment that you have used before.

When asked to make a plan you will often be given some information about the experimental situation to give you a start. You will be expected to apply what you have learned in the theory part of the course. You may also be expected to draw on other sources of information including textbooks, reference books and websites. The emphasis is on making use of the knowledge and experience gained during your course so far.

Working on your own

You will practise planning with the help of your teacher but when it comes to assessment you must work independently to get top marks. Working independently does not mean working in isolation so you should expect to consult sources of reference.

Remember that it can be in your interest to ask for help from your teacher if you are really stuck. If the help gets you started in the right direction, then you can gain credit for all the other aspects of planning. Overall this may allow you to gain more marks. Getting the marks for independent working will often fail to compensate for coming up with a bad plan.

Showing that you have researched and understood the situation

You must make sure that you understand the chemical theory related to the experiment you are going to carry out. You will find a summary of the main ideas you need to be able to apply in Chapters 2–9 in this book.

This book is one of the sources of information you can refer to in your report. Other sources of information, data and support are given in Chapter 13.

In many cases it will help to carry out some quick preliminary experiments to get a feeling for the chemicals and reactions that will be involved in your plan. This can help you to specify the quantities of chemicals to use and the type and scale of the apparatus.

> **Hint**
>
> Give full and accurate references in your plan to all sources of information that you consult (see Chapter 13).

Describing the procedure in detail

Aim to write your plan for an experiment in a way that could be followed by someone else with similar expertise to yourself. Model your written plan on the style of instructions and guidance in practical books or on worksheets from your teacher.

> **Hint**
>
> Remember that a fully labelled diagram, or series of diagrams, can help to make your plan much clearer.

Attending to health and safety

You must carry out a risk assessment and show that you have consulted the relevant guidance on health and safety. See Chapter 11 for details.

1.3 Implementing

In your course you will need time to practise the skills of observation, preparation and analysis before you can be assessed. You will find more detailed guidance on specific practical methods in Chapters 2–9 in this book.

Working unaided in a healthy and safe way

Your teacher will be looking to see that you work cleanly and methodically paying proper attention to health and safety precautions. When learning new skills you may work as a part of a group but you will be on your own when being assessed.

Only carry out practical work in the presence of a qualified supervisor.

Demonstrating practical skill

You will be judged by your results which will depend on the type of experiment you are conducting. Your teacher may make some judgements based on watching you at work in the laboratory. More significant are the outcomes of your work such as the yield and purity of products, your measurements of melting or boiling points, the detail and relevance of your observations and the accuracy of masses, volumes, temperatures and times.

Recording observations and measurements

In your report you must record measurements with precision that matches the quality of the equipment that you have used. Masses should be quoted to the number of decimal places that match the accuracy of the balance you have used. Similarly, with grade B 50 cm^3 burettes, you should note readings to the nearest $0.05\ cm^3$.

1.4 Analysing evidence and drawing conclusions

This is the stage at which you work out the meaning of your observations and measurements. Here you take the raw data and present it in ways that show what you have found out as a result of your practical work.

Forms of communication

You should choose ways of analysing and communicating your findings which are appropriate to your experiment. Consider including labelled diagrams, tables, charts and graphs as well as continuous prose.

You will gain credit for using the language of chemistry correctly. Compare what you write with what you find in textbooks or on websites to check that you are working at the right level for an advanced course.

Processing the data

The later chapters in this book show how you should process the data in quantitative investigations to answer the chemical questions: 'How much?', 'How fast?' and 'How far?'

Your analysis may involve calculations. If so, you must show that you understand each step of the calculation. Always check that the number of significant figures in your final answer is consistent with the precision of your measurements.

Analysis of measurement uncertainty

In any quantitative investigation you will need to assess the measurement uncertainty as described in Chapter 12.

Drawing conclusions

Here you have to relate your findings to your knowledge and understanding of chemical theory. You need to convince your teacher, or an examiner, that you understand the chemical nature of the observations or measurements you are analysing (see Chapters 2–9 for further details).

1.5 Evaluating evidence and procedures

When you evaluate your work you have to make judgements. You might start by making a short list of the criteria that you would use to decide whether or not your practical work has been successful.

Your evaluation should have two parts:

- first you should decide whether or not any results based on measurements are reliable and meaningful, or whether the outcomes of a preparation are adequate considering the methods used

- second you should review the practical techniques that you have used and decide whether they were the right ones to use; this should include comments on any results that seem anomalous.

Commenting on the reliability of data

In a quantitative investigation your analysis should include an estimate of the overall uncertainty in your results.

Comment here on any anomalous results that do not seem to fit in with the rest of your measurements or observations. You should suggest an explanation for any anomalies that you have detected.

Comparing outcomes with expectations

In some investigations you may be able to refer to published data in reference books, or other sources, which allow you to compare your findings with the generally accepted results.

Published descriptions of chemical preparations often quote the likely yield. This allows you to compare your percentage yield with the outcome that can be reasonably expected.

Sometimes the expectations are predictions based on theory. If so, you should discuss this whether or not you judge that your findings are consistent with your predictions.

Identifying possible improvements

You should consider whether or not the method you used could have been improved, either by making minor modifications or by using a completely different approach. Here you can also refer to methods that you could not carry out with the usual range of equipment available in advanced chemistry laboratories. You might refer to other options such as the use of instrumental methods of analysis including the various types of spectroscopy.

> **Hint**
>
> Ask for some help if you are stuck. It is better to lose a mark or two for assistance rather than ending up with nothing because all your work is misguided.

CHAPTER TWO

Chemical amounts

2.1 Principles

Answering the question 'How much?' is central to both chemical analysis and chemical synthesis.

The purpose of quantitative analysis is to answer the question 'How much?' by determining the quantities of chemicals in a sample. In advanced chemistry the main method of quantitative analysis is a titration. Titrations provide a precise way to determine the concentrations of solutions and to investigate the quantities of chemicals involved in reactions.

Chemists also have to answer the question 'How much?' to decide on the quantities of each reactant to add to a reaction mixture and to predict the expected yield of product from a chemical synthesis.

Chemical amounts

When answering the question 'How much?', chemists need to measure amounts of chemicals which contain equal numbers of atoms, molecules or ions. The unit of chemical amount is the mole. One mole is the amount of substance that contains as many specified atoms, molecules or ions as there are atoms in exactly 12 g of carbon-12.

The key to working with chemical amounts in moles is to know the relative masses of atoms on the carbon-12 scale.

The molar mass of an element is numerically equal to the relative mass of the element. The relative mass of carbon is 12. The molar mass of carbon atoms is $12 \, \text{g mol}^{-1}$.

The molar mass of an element or compound is found by adding up the molar masses of the elements in the given formula. The formula of sulfuric acid is H_2SO_4. Its molar mass

$$= (2 \times 1 \, \text{g mol}^{-1}) + (1 \times 32 \, \text{g mol}^{-1}) + (4 \times 16 \, \text{g mol}^{-1})$$

$$= 98 \, \text{g mol}^{-1}$$

> **Note**
>
> Every physical quantity in science has a name, a symbol and a unit. In the case of *amount of substance* the name of the unit is 'mole', the symbol is n and the unit is 'mol'.

Quantities and units

Chemists have balances for determining mass in grams or kilograms. They have graduated glassware for measuring the volumes of liquids and gases. There is, however, no simple measuring instrument for determining chemical amounts directly. Instead chemists first measure masses or volumes and then calculate the chemical amount.

Masses of chemicals and amounts in moles

The most direct way to find the amount of substance in a sample is to weigh it on a balance:

$$\text{amount of substance/mol} = \frac{\text{mass of sample/g}}{\text{molar mass/g mol}^{-1}}$$

This relationship rearranges to give:

$$\text{mass of sample/g} = \text{amount of substance/mol} \times \text{molar mass/g mol}^{-1}$$

Quantities of gases

If the pressure and temperature are fixed, then the volume of a gas depends only on the amount of gas in moles. In other words the volume of a gas is determined by the number of gas molecules present.

The law applies so long as the molecules of a gas are so far apart that their volume is insignificant compared to the volume of the gas and so long as the intermolecular forces can be ignored. For many common gases these criteria apply under normal laboratory conditions. Under conditions where the gas is close to liquefying, the simple rules do not apply.

Avogadro's law follows from the ideal gas equation:

$$pV = nRT \text{ where } R \text{ is the gas constant}$$

If p and T are constant, the volume is proportional to n, the amount of gas in moles.

The molar volume of a gas is the volume of 1 mol of the gas. When $n = 1$:

$$V = \frac{RT}{p}$$

Substituting in this relationship gives values for the molar volume of any gas which behaves like an ideal gas. Two sets of conditions are commonly used for comparing amounts of gases:

- For accurate work the quantities are calculated for standard temperature and pressure, s.t.p. The standard temperature for gases is 273 K and the standard pressure is 101.3 kPa (1 atmosphere). Under these conditions the molar volume of a gas is 22 400 cm³.

- For approximate work, when making estimates under laboratory conditions it is often convenient to use the fact that the molar volume of a gas at around 20 °C and 1 atmosphere pressure is about 24 000 cm³

$$\text{amount of gas/mol} = \frac{\text{volume of gas/cm}^3}{\text{molar volume/cm}^3 \text{ mol}^{-1}}$$

Quantities in solution

Chemists measure the concentrations of solutions in moles per litre:

$$\text{concentration/mol dm}^{-3} = \frac{\text{amount of solute/mol}}{\text{volume of solution/dm}^3}$$

This rearranges to give:

$$\text{amount of solute/mol} = \text{volume of solution/dm}^3 \times \text{concentration/mol dm}^{-3}$$

> **Hint**
>
> Include the units in relationships and use them to check that the units on both sides of the equals sign are consistent.

> **Note**
>
> Avogadro's law states that equal volumes of gases contain equal amounts of gas molecules, in moles, under the same conditions of temperature and pressure.

> **Note**
>
> A litre is a cubic decimetre, dm³.
>
> 1 dm³ = 1000 cm³

2.2 Finding a formula

Purposes

The aim is to find the formula of a hydrated salt. This might, for example, involve finding the value of x in the formula of salts such as hydrated barium sulfate, $BaSO_4.xH_2O$ or hydrated iron(II) sulfate, $FeSO_4.xH_2O$. This can be done by heating a measured sample of the salt to dryness and finding the loss in mass. The method is only suitable if the anhydrous salt itself does not decompose on heating.

Method

⚠

Check safety before carrying out any practical procedure.

Figure 2.1 below shows the outline procedure. A lid is usually necessary in the early stages of heating to prevent the loss of crystal fragments while the bulk of the water is being driven off.

Step 1 Weigh a crucible and lid. Place the planned quantity of the hydrated salt in the crucible and reweigh the crucible, lid and hydrated salt.

(a)

Step 2 Heat the crucible and its contents for about 10 minutes. Start heating gently and then more strongly. During the last minute of heating remove the cover so that any moisture which has collected on the underside of the cover can evaporate.

(b)

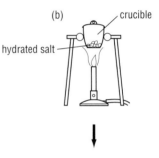

Step 4 Replace the cover and heat the crucible for 5 more minutes, removing the cover during the last minute of heating as before. Cool and reweigh the crucible, lid and contents. This last mass should agree with the previous mass to within the accuracy of the balance. If it does not, repeat this heating until a constant mass is reached.

(c)

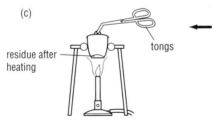

Step 3 Allow the crucible to cool. Then reweigh the crucible, lid and contents.

Figure 2.1 A procedure for finding the mass of water of crystallisation in a sample of a hydrated salt

Steps 3 and 4 are repeated until the results show that all the water of crystallisation has been driven off. Once this has happened the combined mass of the crucible, lid and residue ceases to fall. Chemists call this 'heating to constant mass'.

Results

Mass of crucible plus lid	= 18.55 g
Mass of crucible, lid and hydrated salt crystals	= 23.95 g
Mass of crucible, lid and residue after heating to constant weight	= 21.49 g

Calculation

Mass of hydrated iron(II) sulfate before heating = 5.40 g

Mass of anhydrous $FeSO_4$ after heating $\qquad$ = 2.94 g

Loss in mass of water $\qquad$ = 2.46 g

	$FeSO_4$	H_2O
Masses combining/g	2.94	2.46
Molar mass/g mol^{-1}	152	18
Amounts combining/mol	$\dfrac{2.94}{152} = 0.0193$	$\dfrac{2.46}{18} = 0.137$
$\dfrac{\text{Amount/mol}}{\text{Smaller amount/mol}}$	$\dfrac{0.0193}{0.0193} = 1.00$	$\dfrac{0.137}{0.0193} = 7.10$
Simplest whole number ratio	1	7

Value of x in $FeSO_4.xH_2O = 7$

Formula of hydrated salt = $FeSO_4.7H_2O$

2.3 Which equation is correct?

Purposes

Chemists can determine the equation for a reaction by measuring the masses of reactants and products. If any of the elements or compounds are gases then their relative amounts can be found by measuring gas volumes.

The aim of the experiment described here is to find out whether it is iron(II) sulfate or iron(III) sulfate that forms when iron displaces copper metal from a solution of copper(II) sulfate (Figure 2.2).

If the product is the iron(II) compound then the equation is:

$$Fe(s) + CuSO_4(aq) \rightarrow FeSO_4(aq) + Cu(s)$$

$\quad$ 1 mol $\qquad\qquad\qquad\qquad$ 1 mol

If, however the product is the iron(III) compound the equation is:

$$2Fe(s) + 3CuSO_4(aq) \rightarrow Fe_2(SO_4)_3(aq) + 3Cu(s)$$

$\quad$ 2 mol $\qquad\qquad\qquad\qquad\qquad$ 3 mol

Check safety before carrying out any practical procedure.

Methods

Step 1 Add about 0.6 g iron filings to an empty basin from a weighing bottle. Then weigh the bottle again.

Step 2 Slowly add a warm solution of copper(II) sulfate and stir.

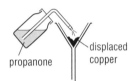
about 2.5 g of copper(II) sulfate in 25 cm³ of warm water

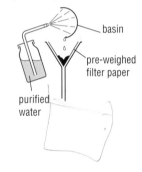
basin
weighed sample of iron filings

Step 5 Open out the filter paper. Allow it to dry in a warm oven. Weigh the filter paper and its contents.

Step 4 Wash the copper on the paper – first with purified water and then with propanone.

Step 3 Weigh a dry filter paper and fold it to fit a funnel. Filter the contents of the basin using purified water to wash all the displaced copper into the funnel.

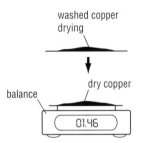
washed copper drying
dry copper
balance
01.46

propanone
displaced copper

basin
pre-weighed filter paper
purified water

Figure 2.2 A procedure for investigating the equation for the reaction of iron with copper(II) sulfate solution

Results

As the reaction occurs the solution becomes hot and the blue copper ion solution becomes paler. The iron dissolves and pink copper powder precipitates.

Mass of weighing bottle plus iron = 4.42 g

Mass of bottle after tipping out the iron = 3.89 g

Mass of dry filter paper = 0.84 g

Mass of paper plus displaced copper = 1.46 g

Calculation

Mass of iron added to excess copper(II) sulfate = 0.53 g

Mass of copper displaced = 0.62 g

Ratio of mass of copper to mass of iron $= \dfrac{0.62\,g}{0.53\,g} = 1.17$

Molar mass of iron $= 56\,g\,mol^{-1}$

Molar mass of copper $= 64\,g\,mol^{-1}$

Expected ratio of masses if the product is an iron(II) compound $= \dfrac{64\,g}{56\,g} = 1.14$

Expected ratio of masses if the product is an iron(III) compound $= \dfrac{(3 \times 64\,g)}{(2 \times 56\,g)} = 1.71$

The results suggest that the reaction produces the iron(II) compound.

2.4 Reacting quantities and yields

Purposes

In designing a procedure for making a chemical product, chemists aim to make each reaction as efficient as possible. Chemists use the balanced equation for the reaction to decide on the quantities of reactants to start with. They also calculate the maximum yield that can be obtained in theory from the starting materials.

One of the reactants is likely to be more expensive or scarcer than the others so it is common to add the other reagents in excess. This helps to convert as much as possible of the more valuable chemical into the desired product.

Calculating yields

The theoretical yield for a synthesis is the mass of product expected assuming that the reaction goes according to the balanced chemical equation.

The actual yield is the amount of product obtained from the synthesis.

The percentage yield is given by this relationship:

$$\text{percentage yield} = \frac{\text{actual yield}}{\text{theoretical yield}} \times 100\%$$

Example

A synthesis of 1-bromobutane produces an actual yield of 6.8 g from 7.5 cm^3 of butan-1-ol (see Section 9.2).

The equation for the reaction shows that excess of sodium bromide and concentrated sulfuric acid converts 1 mol butan-1-ol to 1 mol 1-bromobutane.

$C_4H_9OH(l) \rightarrow C_4H_9Br(l)$

Molar mass of butan-1-ol	$= (4 \times 12) + (10 \times 1) = 74 \text{ g mol}^{-1}$
Molar mass of 1-bromobutane	$= (4 \times 12) + (9 \times 1) + 80 = 137 \text{ g mol}^{-1}$

So theoretically the yield from 74 g butan-1-ol should be 137 g of 1-bromobutane.

The density of butan-1-ol is 0.81 g cm^{-3}, so the mass of alcohol used	$= 7.5 \text{ cm}^3 \times 0.81 \text{ g cm}^{-3} = 6.1 \text{ g}$
The theoretical yield of 1-bromobutane	$= 6.1 \text{ g} \times \dfrac{137}{74} = 11.3 \text{ g}$
The actual yield was 6.8 g, so the percentage yield	$= \dfrac{6.8 \text{ g}}{11.3 \text{ g}} \times 100\% = 60\%$

Planning a synthesis

The synthesis of the drug antifebrin described in Section 9.3 shows the procedure for converting phenylammonium chloride into the product.

Adding a solution of sodium ethanoate converts the amine salt into phenylamine. The phenylamine then reacts with ethanoic anhydride to produce the antifebrin which is insoluble in cold water and separates as a solid.

Formation of phenylamine from its salt:

$$C_6H_5NH_3{}^+Cl^-(aq) + CH_3COO^-Na^+(aq)$$
$$\rightarrow C_6H_5NH_2(aq) + CH_3COOH(aq) + Na^+Cl^-(aq)$$

Acylation with ethanoic anhydride:

$$C_6H_5NH_2(aq) + (CH_3CO)_2O(aq) \rightarrow CH_3CONHC_6H_5(s) + CH_3COOH(aq)$$

The equations show that 1 mol of $C_6H_5NH_3{}^+Cl^-$ produces 1 mol of the required product, $CH_3CONHC_6H_5(s)$. In planning this synthesis it makes sense to use excess of the two reagents used: sodium ethanoate and ethanoic anhydride.

Estimating a suitable quantity of phenylammonium chloride
Suppose that the aim is to make 1.0 g of the product.

The molar mass of phenylammonium chloride $= 129.5\,\text{g mol}^{-1}$

The molar mass of the product, antifebrin $\quad = 125\,\text{g mol}^{-1}$

Experience suggests that after all stages of the preparation and purification of the product a percentage yield of about 40 % is likely. So to produce 1.0 g of product the theoretical yield must be:

$$\frac{100}{40} \times 1.0\,\text{g} \quad = 2.5\,\text{g}$$

The mass of phenylammonium chloride required to give a theoretical yield of 2.5 g is:

$$\frac{129.5}{125} \times 2.5\,\text{g} = 2.6\,\text{g}$$

So in practice it would make sense to start with 3.0 g of the solid since the yield may not in fact be quite as high as predicted.

Estimating the quantities of the other reagents
The chemical amount of phenylammonium chloride in 3.0 g of the solid
$$= 0.023\,\text{mol}$$

In theory this would react with 0.023 mol of sodium ethanoate crystals or, as the molar mass of the hydrated salt is $136\,\text{g mol}^{-1}$, with:

$$= 0.023\,\text{mol} \times 136\,\text{g mol}^{-1} = 3.13\,\text{g}$$

However the sodium ethanoate should be added in considerable excess to convert all the salt to the free amine. In practice a solution of about 18.0 g of the salt in water might be used. This figure cannot easily be predicted and comes from experience based partly on trial and error.

Similarly, in theory, 0.023 mol of ethanoic anhydride (molar mass $= 102$ g mol^{-1}) is needed for the next step.

$$0.023 \text{ mol} \times 102 \text{ g mol}^{-1} = 2.35 \text{ g}$$

Here experience shows that it is satisfactory to use slightly more than double the required quantity of the acylating agent. Since the density of ethanoic anhydride is 1.1 g cm^{-3}, this means using about 6 cm^3 of the liquid reagent.

2.5 Practical skills

Planning

You will have to make estimates of quantities when planning a wide range of quantitative investigations.

Once you have decided on the practical approach to your investigation you can carry out calculations to estimate the masses or volumes expected from the relevant balanced equations and then choose quantities which are suitable for standard laboratory apparatus. You should show the details of your calculations and specify the capacities of the equipment you are going to use.

When planning to prepare a sample of a compound, you need to use the equation for the reaction to calculate the theoretical mass of the limiting starting chemical needed to make the specified quantity of product. You then have to work out how much of this chemical you actually need to start with, allowing for the fact that in practice the percentage yield will be less than 100 per cent.

Always show your plan to your teacher before starting any practical work.

Note

The limiting reagent is the chemical in a reaction mixture which limits the theoretical yield. In many chemical preparations the other reactants are added in excess.

Test Yourself

2.1 A student plans to measure the rate of reaction of magnesium ribbon with dilute hydrochloric acid by collecting the gas formed in a gas syringe that can contain up to 100 cm^3 gas.

 (a) What is the maximum mass of magnesium that can be used in the experiment?

 (b) Is it practicable to work with the mass of metal calculated in **(a)**?

2.2 What mass of magnesium carbonate reacts with excess of dilute sulfuric acid to give a solution from which, in theory, it would be possible to form 10 g of crystals of the hydrated salt MgSO$_4$.7H$_2$O?

2.3 The liquid alkene, 2-methylbut-2-ene, reacts with slight excess of bromine to produce 2,3-dibromo-2-methylbutane. The expected percentage yield after separating and purifying the product is about 80%. What volumes of the alkene and bromine would you use to make 20 g of the product? (The density of the alkene $= 0.66$ g cm^{-3} and the density of bromine $= 3.12$ g cm^{-3}.)

Analysing and drawing conclusions

As you make measurements in a quantitative experiment it is important to record all your readings neatly, including the units. You will often summarise your measurements in a table. You may then need to plot a graph to display and interpret your results.

Your analysis should include an estimate of the percentage uncertainty (error) in each critical measurement. You also need to include the maximum total uncertainty in values you have calculated in your results (see Chapter 12).

Evaluation

In your evaluation of any experiment where you have measured quantities of chemicals you should consider commenting on:

- the consistency of measurements where you have carried out the measurements with two or more samples

- whether or not your conclusions can be relied on in view of the total estimated uncertainty (error) in your calculated results

- the percentage difference between the value you have arrived at and the value expected according to theory or books of data

- ways by which the procedure and methods of taking measurements could be improved.

Test Yourself

2.4 On heating to constant mass, a 7.30 g sample of a hydrate of magnesium bromide, $MgBr_2.xH_2O$, left a residue of 4.60 g of the anhydrous salt. What is the formula of the hydrate?

2.5 A small pellet of lithium was rinsed in an organic solvent to remove the oil in which it was stored. The pellet was quickly dried between filter papers and weighed. The dry pellet was added to water and the volume of hydrogen produced was collected and measured at room temperature and pressure. The mass of the pellet was 0.021 g and the volume of gas collected was 35 cm^3.

(a) Show that these results are consistent with the balanced equation if the metal reacts with water to form lithium hydroxide, LiOH.

(b) Suggest reasons why the measured volume of gas was not exactly as predicted by the equation.

2.6 Calculate the theoretical and percentage yields if a preparation produces 6.2 g of the ester ethyl ethanoate, $CH_3COOC_2H_5$, from 5.0 g ethanol reacting with excess ethanoic acid.

CHAPTER THREE

Volumetric analysis

Principles

Accurate chemical analysis often involves preparing a solution of an unknown sample. Then it may be necessary to dilute a solution quantitatively. Next the analysis may require some form of titration to measure the volume of the sample solution that reacts with a certain volume of a reference solution with an accurately known concentration.

Today many laboratories have automatic instruments for carrying out titrations, but the principle is exactly the same as for a titration where the volumes are measured with a traditional burette and pipette. Volumetric titrations with the kinds of glassware seen in school and college laboratories continue to be widely used in the food, pharmaceutical and other industries.

Any titration involves two solutions. Typically a measured volume of one solution, from a pipette, is run into a flask. Then the second solution is added bit by bit from a burette until the colour change of an indicator, or changing signal from an instrument, shows that the reaction is complete.

The procedure only gives accurate results if the reaction between the two solutions is rapid and proceeds exactly as described by the chemical equation. So long as these conditions apply, titrations can be used to study acid–base, redox, precipitation and complex-forming reactions.

> **Note**
>
> A reaction is stoichiometric if it proceeds exactly as in the balanced equation. The reaction in a titration must be stoichiometric, fast and go to completion.

3.2 Procedures

Preparing a standard solution

Standard solutions make accurate volumetric analysis possible. A standard solution is any solution with an accurately known concentration. The direct way of preparing a standard solution is to dissolve a known mass of a chemical in water and then to make the volume of solution up to a definite volume in a graduated flask (Figure 3.1).

Figure 3.1 *The procedure for preparing a standard solution from a primary standard*

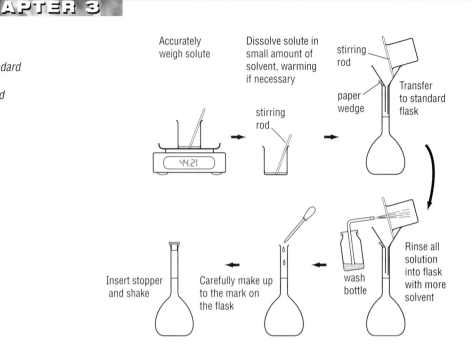

Accurately weigh solute

Dissolve solute in small amount of solvent, warming if necessary

stirring rod

stirring rod

paper wedge

Transfer to standard flask

Insert stopper and shake

Carefully make up to the mark on the flask

wash bottle

Rinse all solution into flask with more solvent

This method for preparing a standard solution is only appropriate with a chemical that:

● is very pure
● does not gain or lose mass when in the air
● has a relatively high molar mass so that weighing errors are minimised.

Chemicals that meet these criteria are primary standards. A titration with a primary standard can then be used to measure the concentration of the solution to be analysed.

Table 3.1 *Examples of primary standards*

Primary standard	Molar mass/g mol⁻¹	Used to standardise (type of titration)
sodium carbonate, anhydrous	106.0	acids (acid–base)
potassium hydrogenphthalate (potassium hydrogenbenzene -1,2- dicarboxylate)	204.2	bases (acid–base)
potassium dichromate(VI)	294.2	reducing agents (redox)
potassium iodate	214.0	sodium thiosulfate (redox)
edta (disodium salt, dihydrate)	372.3	metal ions (complex-forming)

Diluting a solution quantitatively

Quantitative dilution is an important procedure in analysis. Two common reasons for carrying out dilutions are to:

- make a solution with known concentration by diluting a standard solution
- dilute an unknown sample for analysis to give a concentration suitable for titration.

Successive dilutions of a standard solution can provide a series of solutions used to calibrate instruments such as colorimeters.

The procedure for dilution is to take a measured volume of the more concentrated solution with a pipette (or burette) and run it into a graduated flask. The flask is then carefully filled to the mark with purified water.

The key to calculating the volumes to use when diluting a solution is to remember that the amount in moles of the chemical dissolved in the diluted solution must equal the amount in moles of the sample taken from the concentrated solution. If c is the concentration in mol dm^{-3} and V is the volume in dm^3, then:

- the amount in moles of the chemical in the measured volume of the concentrated solution $= c_A V_A$
- the amount in moles of the chemical in the diluted solution $= c_B V_B$

These two amounts are the same and so: $c_A V_A = c_B V_B$

Use a safety filler whenever you use a pipette.

Example

You require a $0.1\ \text{mol dm}^{-3}$ solution of NaOH(aq). You are supplied with a $250\ \text{cm}^3$ graduated flask and a supply of $0.5\ \text{mol dm}^{-3}$ sodium hydroxide solution. What volume of the concentrated solution will you measure into the graduated flask?

Answer

$c_A = 0.5\ \text{mol dm}^{-3}$ $\qquad\qquad$ $c_B = 0.1\ \text{mol dm}^{-3}$

$V_A = $ to be calculated $\qquad\quad$ $V_B = 250\ \text{cm}^3 = 0.25\ \text{dm}^3$

$$V_A = \frac{0.1\ \text{mol dm}^{-3} \times 0.25\ \text{dm}^3}{0.5\ \text{mol dm}^{-3}} = 0.05\ \text{dm}^3 = 50\ \text{cm}^3$$

Pipetting $50\ \text{cm}^3$ of the concentrated solution into the $250\ \text{cm}^3$ graduated flask and making up to the mark with pure water gives the required dilution after thorough mixing.

Note

$1\ \text{dm}^3 = 1000\ \text{cm}^3$
$\qquad\quad = 1\ \text{litre}$

Hint

Check dilution calculations by inspection. In the example the solution has to be diluted by a factor of five, so the final volume of the diluted solution has to be five times the volume of concentrated solution run into the flask.

Titrations

A titration involves two solutions. A measured volume of one solution is run into a flask. The second solution is then added bit by bit from a burette until the reaction is complete.

Some titrations are used to investigate reactions. In these experiments the concentrations of both solutions are known and the aim is to determine the equation for the reaction.

More often in titrations, the purpose is to measure the concentration of an unknown solution given the equation for the reaction and using a second solution of known concentration.

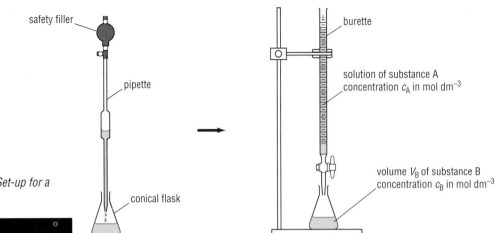

Figure 3.2 Set-up for a titration

safety filler

pipette

conical flask

burette

solution of substance A concentration c_A in mol dm^{-3}

volume V_B of substance B concentration c_B in mol dm^{-3}

Figure 3.2 shows the apparatus used for a titration based on a reaction between two chemicals A and B. The situation is essentially the same for acid–base, redox and complex forming titrations.

$$n_A A + n_B B \rightarrow \text{products}$$

This means that n_A moles of A react with n_B moles of B.

The concentration of solution B in the flask is c_B measured in mol dm^{-3}. The concentration of solution A in the burette is c_A measured in mol dm^{-3}.

The analyst uses a pipette to run a volume V_B of solution B into the flask. Then the analyst adds solution A from the burette until an indicator shows that the reaction is complete. At the end-point, the volume added is the titre, V_A. The analyst repeats the titration enough times to achieve consistent results (see page 28).

In the laboratory, volumes of solutions are normally measured in cm^3 but they should be converted to dm^3 in calculations so that they are consistent with the units to measure concentrations.

The amount in moles of B in the flask at the start $= c_B \times V_B$

The amount in moles of A added from the burette $= c_A \times V_A$

The ratio of these amounts must be the same as the ratio of the amounts shown in the equation:

$$\frac{c_A \times V_A}{c_B \times V_B} = \frac{n_A}{n_B}$$

In any titration, all but one of the values in this relationship are known. The one unknown is calculated from the results.

In titrations to analyse solutions, the equation for the reaction is given so that the ratio n_A/n_B is known. The concentration of one of the solutions is also known. The volumes V_A and V_B are measured during the titration. Substituting all the known quantities in the titration formula allows the concentration of the unknown solution to be calculated.

In titrations to investigate reactions, the problem is to determine the ratio n_A/n_B. The concentrations c_A and c_B are known and the volumes V_A and V_B are measured during the titration. So the desired ratio can be calculated from the formula.

3.3 Acid–base titrations

Determining the solubility of calcium hydroxide

Calcium hydroxide is an alkali which is only slightly soluble. Its solubility can be determined by titration of a saturated solution of the alkali with a standard solution of hydrochloric acid. The equation for the reaction is known, so the unknown to be determined is the concentration of the saturated solution of the alkali.

The results of the titration are shown in Figure 3.3.

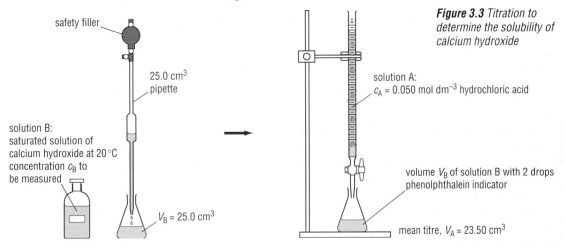

safety filler

25.0 cm³ pipette

solution B:
saturated solution of calcium hydroxide at 20 °C concentration c_B to be measured

V_B = 25.0 cm³

Figure 3.3 Titration to determine the solubility of calcium hydroxide

solution A:
c_A = 0.050 mol dm⁻³ hydrochloric acid

volume V_B of solution B with 2 drops phenolphthalein indicator

mean titre, V_A = 23.50 cm³

The equation for the titration reaction is:

$$Ca(OH)_2(aq) + 2HCl(aq) \rightarrow CaCl_2(aq) + 2H_2O(l)$$

So 1 mol of the alkali reacts with 2 mol of the acid.

Substituting the values in the titration formula gives:

$$\frac{0.050 \times 23.50}{c_B \times 25.00} = \frac{2}{1}$$

Rearranging, this, becomes:

$$c_B = \frac{0.050 \times 23.50}{2 \times 25.00} = 0.0235 \text{ mol dm}^{-3}$$

So the concentration of saturated limewater = 0.0235 mol dm⁻³

The molar mass of calcium hydroxide = 74.1 g mol⁻¹

So the concentration of saturated limewater = 0.0235 mol dm⁻³ × 74.1 g mol⁻¹

$$= 1.74 \text{ g dm}^{-3}$$

19

Finding the percentage of calcium carbonate in egg shell

Calcium carbonate reacts with acids but it is insoluble and reacts too slowly for a direct titration. It has to be determined by a procedure called a 'back titration'.

The analyst adds an excess of a standard solution of hydrochloric acid to a weighed sample of the crushed shell. Then the results of a titration make it possible to measure the amount of excess hydrochloric acid and hence the amount that must have reacted with $CaCO_3$.

Thus, when the reaction is complete the next step is to transfer the solution quantitatively to a graduated flask. Water is added to the mark and the diluted solution is well mixed. Finally the analyst titrates measured volumes of the solution with a standard solution of sodium hydroxide to determine the amount of unreacted acid (Figure 3.4).

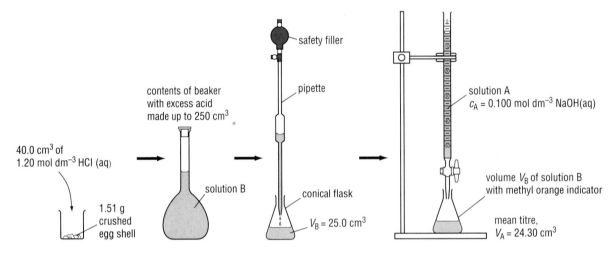

Figure 3.4 Determining the percentage of calcium carbonate in egg shells

In this analysis there are two reactions. First some of the added hydrochloric acid reacts with the egg shells. Then, in the titration, a portion of the excess acid reacts with the sodium hydroxide. The titration makes it possible to find out by how much the acid was in excess.

Calculating the amount of excess acid

The equation for the titration reaction is:

$$HCl(aq) + NaOH(aq) \rightarrow NaCl(aq) + H_2O(l)$$

So 1 mol of the HCl reacts with 1 mol of NaOH.

Substituting the values in the titration formula gives:

$$\frac{0.100 \times 24.30}{c_B \times 25.0} = \frac{1}{1}$$

Rearranging, this becomes:

$$c_B = \frac{0.100 \times 24.30}{25.0} = 0.0972 \text{ mol dm}^{-3}$$

The concentration of excess acid in the graduated flask

$$= 0.0972 \text{ mol dm}^{-3}$$

The volume of the graduated flask	$= 250 \text{ cm}^3 = 0.25 \text{ dm}^3$
So the total amount of excess HCl	$= 0.25 \text{ dm}^3 \times 0.0972 \text{ mol dm}^{-3}$
	$= 0.0243 \text{ mol}$

Calculating the amount of calcium carbonate in the egg shells

The amount of acid added to the egg shells $= 0.040 \text{ dm}^3 \times 1.20 \text{ mol dm}^{-3}$
$$= 0.0480 \text{ mol}$$

Amount of HCl that reacted with egg shell $= 0.0480 \text{ mol} - 0.0243 \text{ mol}$
$$= 0.0237 \text{ mol}$$

The equation for the reaction of the acid with the egg shells:

$$CaCO_3(s) + 2HCl(aq) \rightarrow CaCl_2(aq) + CO_2(g) + H_2O(l)$$

So the amount in moles of $CaCO_3$ in 1.51 g egg shells is half the amount in moles of HCl that reacted $= 0.0118 \text{ mol}$.

The molar mass of $CaCO_3$	$= 100 \text{ g mol}^{-1}$
So the mass of $CaCO_3$ in 1.51 g egg shells	$= 0.0118 \text{ mol} \times 100 \text{ g mol}^{-1}$
	$= 1.18 \text{ g}$

The percentage of $CaCO_3$ in the egg shells $= \dfrac{1.18 \text{ g}}{1.51 \text{ g}} \times 100\% = 78.1\%$

3.4 Redox titrations

Finding the value of x in Fe(NH₄)₂(SO₄)₂.xH₂O

Potassium manganate(VII), $KMnO_4$, reacts quantitatively with many reducing agents including iron(II) ions. When acting as an oxidising agent in acid solution the deep purple MnO_4^- ions turn to almost colourless Mn^{2+} ions. No added indicator is required for these titrations. A permanent pink colour on adding one more drop of the aqueous potassium manganate(VII) indicates the end-point.

It is possible to determine x in the formula $FeSO_4(NH_4)_2SO_4.xH_2O$ by making up a solution of the salt with a known concentration in grams per litre and then using a titration with potassium manganate(VII) to find the concentration in moles per litre. The results make it possible to calculate the molar mass of the salt and hence the value of x in the formula (Figure 3.5).

> **Hint**
>
> Potassium manganate(VII) is so intensely coloured that the bottom of the meniscus may be invisible in a burette. This is the one time to take readings from the top of the meniscus.

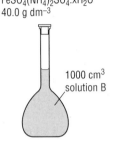

solution of
$FeSO_4(NH_4)_2SO_4.xH_2O$
40.0 g dm^{-3}

1000 cm^3
solution B

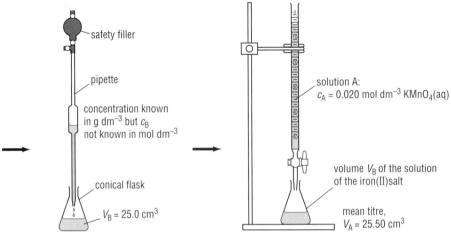

safety filler

pipette

concentration known
in g dm^{-3} but c_B
not known in mol dm^{-3}

conical flask

V_B = 25.0 cm^3

solution A:
c_A = 0.020 mol dm^{-3} KMnO$_4$(aq)

volume V_B of the solution
of the iron(II) salt

mean titre,
V_A = 25.50 cm^3

Figure 3.5 *Finding the formula of ammonium iron(II) sulfate*

When manganate(VII) ions react in acid solution the half-equation is:

$$MnO_4^-(aq) + 8H^+(aq) + 5e^- \rightarrow Mn^{2+}(aq) + 4H_2O(l)$$

While the half-equation for the oxidation of iron(II) is:

$$Fe^{2+}(aq) \rightarrow Fe^{3+}(aq) + e^-$$

So 1 mol of MnO_4^- oxidises 5 mol of Fe^{2+}.

Substituting the values in the titration formula gives:

$$\frac{0.020 \times 25.50}{c_B \times 25.0} = \frac{1}{5}$$

Rearranging, this becomes:

$$c_B = \frac{0.020 \times 25.50 \times 5}{25.0} = 0.102 \text{ mol dm}^{-3}$$

Concentration of iron(II) ions in the graduated flask = 0.102 mol dm^{-3}

So a solution containing 40.0 g dm^{-3} of $FeSO_4(NH_4)_2SO_4.xH_2O$ contains 0.102 mol dm^{-3} of iron(II) ions.

One mole of iron(II) is present in one mole of the compound, therefore:

$$\text{molar mass of } FeSO_4(NH_4)_2SO_4.xH_2O = \left(\frac{40.0 \text{ g dm}^{-3}}{0.102 \text{ mol dm}^{-3}}\right) = 392.2 \text{ g mol}^{-1}$$

The molar mass of $FeSO_4(NH_4)_2SO_4$ = 284.0 g mol^{-1}

The difference of 108.2 g mol^{-1} is accounted for by the water of crystallisation.

The molar mass of H_2O = 18 g mol^{-1}.

$$\text{Hence the value of x} = \frac{108.2 \text{ g mol}^{-1}}{18 \text{ g mol}^{-1}} = 6.0$$

The formula of the salt is $FeSO_4(NH_4)_2SO_4.6H_2O$.

Finding the concentration of chlorine bleach

Thiosulfate ions react quantitatively with iodine. The thiosulfate ions reduce the iodine to iodide:

$$2S_2O_3^{2-}(aq) \rightarrow S_4O_6^{2-}(aq) + 2e^-$$

$$I_2(aq) + 2e^- \rightarrow 2I^-(aq)$$

These two half-equations show that 2 mol thiosulfate ion reduces 1 mol iodine molecules.

This system can be used to investigate quantitatively any oxidising agent that can oxidise iodide ions to iodine. The procedure is to:

- add excess potassium iodide to a measured quantity of the oxidising agent which then converts iodide ion to iodine

- titrate the iodine formed with a standard solution of sodium thiosulfate.

At the end-point the iodine colour disappears. This is a change from pale yellow to colourless. Adding a few drops of starch solution just before the end-point makes the colour change much sharper. Starch gives a deep blue colour with iodine which disappears at the end-point.

Domestic chlorine bleach is a solution of sodium chlorate(I). If the solution is acidified it produces chlorine. The amount of chlorine produced from a measured quantity of bleach is the 'available' chlorine.

The available chlorine from bleach can be estimated by adding excess potassium iodide and then titrating with sodium thiosulfate (Figure 3.6). The bleach has to be diluted quantitatively to give a reasonable titre with sodium thiosulfate with a concentration around $0.1 \, \text{mol dm}^{-3}$.

> **Note**
>
> Iodine is only very sparingly soluble in water. It dissolves much more readily in excess potassium iodide solution. Iodine reacts reversibly with iodide ions giving a solution which is dark brown when more concentrated but pale yellow when dilute.

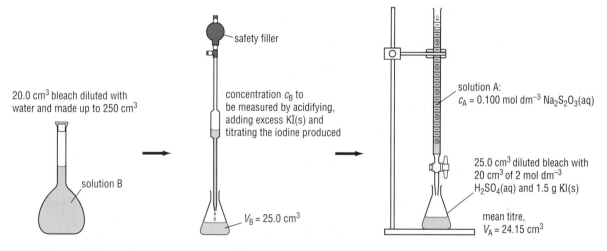

20.0 cm³ bleach diluted with water and made up to 250 cm³

solution B

safety filler

concentration c_B to be measured by acidifying, adding excess KI(s) and titrating the iodine produced

$V_B = 25.0 \, \text{cm}^3$

solution A:
$c_A = 0.100 \, \text{mol dm}^{-3} \, \text{Na}_2\text{S}_2\text{O}_3(aq)$

25.0 cm³ diluted bleach with 20 cm³ of 2 mol dm⁻³ $H_2SO_4(aq)$ and 1.5 g KI(s)

mean titre,
$V_A = 24.15 \, \text{cm}^3$

Figure 3.6 *Measuring the concentration of chlorine bleach*

The equation for the titration reaction is:

$$2S_2O_3^{2-}(aq) + I_2(aq) \rightarrow S_4O_6^{2-}(aq) + 2I^-(aq)$$

Note

When using
thiosulfate to titrate
iodine formed by an
oxidising agent,
iodide ions are first
oxidised to iodine
and then reduced
back to iodide. So
you do not need to
include the iodine in
the calculation. It is
present as a 'go
between'. Relate the
thiosulfate directly
to the oxidising
agent with the help
of the equations.

The iodine was formed by oxidation of iodide ions by bleach:

$$ClO^-(aq) + 2H^+(aq) + 2I^-(aq) \rightarrow I_2(aq) + H_2O(l) + Cl^-(aq)$$

So 2 mol of thiosulfate ions reacts with 1 mol of iodine which in turn was produced by reaction with 1 mol chlorate(I). The ratio of moles thiosulfate to moles chlorate(I) is 2:1.

Substituting the values in the titration formula gives:

$$\frac{0.10 \times 24.15}{c_B \times 25.00} = \frac{2}{1}$$

Rearranging, this becomes:

$$c_B = \frac{0.10 \times 24.15}{2 \times 25.00} = 0.0483 \text{ mol dm}^{-3}$$

So the concentration of chlorate(I) ions in diluted bleach = 0.0483 mol dm^{-3}

20.0 cm^3 of the original bleach was diluted to 250 cm^3, so the concentration of chlorate(I) ions in the concentrated bleach

$$= \frac{250}{20} \times 0.0483 \text{ mol dm}^{-3} = 0.604 \text{ mol dm}^{-3}$$

The 'available chlorine' in chlorate bleach is set free as the element on adding acid according to this equation, which shows that 1 mol chlorate(I) reacts with chloride ions to produce 1 mol chlorine.

$$ClO^-(aq) + 2H^+(aq) + Cl^-(aq)) \rightarrow Cl_2(aq) + H_2O(l)$$

So the 'available chlorine' in concentrated bleach   = 0.604 mol dm^{-3}

The molar mass of chlorine molecules   = 71.0 g mol^{-1}

So the 'available chlorine' in the concentrated bleach = 42.9 g dm^{-3}

3.5   Complex-forming titrations

Note

The initials edta
derive from the
traditional name
ethylene diamine
tetra acetic acid.

The hexadentate ligand, edta, forms very stable complexes with metal ions. The complex-forming reaction is rapid and quantitative so it can be used to measure the concentrations of ions such as magnesium and calcium ions in solution.

The ligand, edta, is supplied as a disodium salt. The salt is hydrated and can be represented as $Na_2H_2Y.2H_2O$. In solution the negative ion reacts with metal ions such as magnesium ions:

$$Mg^{2+}(aq) + H_2Y^{2-}(aq) \rightarrow MgY^{2-}(aq) + 2H^+(aq)$$

The end-point can be detected by indicators which form coloured complexes with metal ions which are less stable than the edta complexes. The pH has to be controlled. A buffer solution is added to the titration flask to keep the solution alkaline.

A suitable indicator is Eriochrome black T which is blue in solution at pH 10. At the start of the titration the indicator forms a wine-red complex with the metal ion present. Edta from the burette forms a more stable complex and so takes the metal ions from the indicator. At the end-point all the metal ions are complexed with edta. The last drop leaves no metal ions to make the red complex with the indicator so the solution turns blue again.

Measuring the concentration of zinc ions in a dietary supplement

Suppliers of mineral and vitamin supplements sell a solution of zinc sulfate. This solution can be used in a test for zinc deficiency or taken as a dietary supplement. The main ingredient is zinc sulfate. The concentration can be estimated by titration with edta (Figure 3.7).

The equation for the titration reaction is:

$$Zn^{2+}(aq) + H_2Y^{2-}(aq) \rightarrow ZnY^{2-}(aq) + 2H^+(aq)$$

So 1 mol of the edta reacts with 1 mol of zinc ions.

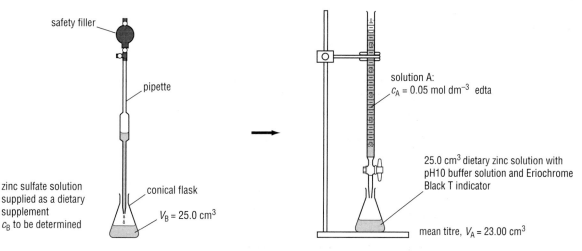

solution A:
$c_A = 0.05$ mol dm^{-3} edta

25.0 cm^3 dietary zinc solution with pH10 buffer solution and Eriochrome Black T indicator

mean titre, $V_A = 23.00$ cm^3

zinc sulfate solution supplied as a dietary supplement c_B to be determined

safety filler

pipette

conical flask

$V_B = 25.0$ cm^3

Substituting the values in the titration formula gives:

$$\frac{0.050 \times 23.00}{c_B \times 25.0} = \frac{1}{1}$$

Figure 3.7 *Estimating the concentration of zinc ions in a dietary supplement*

Rearranging, this becomes:

$$c_B = \frac{0.050 \times 23.00}{25.0} = 0.0460 \text{ mol dm}^{-3}$$

So the concentration of zinc ions $= 0.0460 \text{ mol dm}^{-3}$

The usual dose of the supplement is 5 cm^3.

The amount of zinc ions in this dose $= \dfrac{5}{1000}$ dm$^3 \times 0.0460$ mol dm^{-3}

$= 0.00023$ mol

The molar mass of Zn^{2+} $= 65.4 \text{ g mol}^{-1}$

So the dose of zinc ions per 5 cm^3 of the supplement

$= 65.4 \text{ g mol}^{-1} \times 0.00023 \text{ mol}$
$= 0.015 \text{ g} = 15 \text{ mg}$

3.6 Practical skills

You are certain to be assessed on your ability to plan, carry out and interpret titrations. All your practical skills will be tested either through coursework or by examination.

Planning

Always show your plan to your teacher before starting any practical work.

Use of references

If you are making your plan as part of your coursework you should consult at least two references (see Chapter 13).

There are many websites offering practical instructions for titrations which you can find with the help of search engines. Try search terms which specify the type of titration (acid–base, redox or complexometric) followed by the name or chemical to be analysed. Bear in mind that some of the procedures you find will not be suitable for an advanced science course in school or college.

Checklist

When planning a titration make sure that you:
- ★ list the equipment, apparatus and chemicals
- ★ specify and justify quantities and concentrations
- ★ give step-by-step instructions that another student could follow successfully
- ★ identify any hazards and how the risk from them can be reduced
- ★ provide a full explanation for each step
- ★ indicate how you would calculate the required value from the experimental results
- ★ identify sources of reference.

Concentrations of solutions and amounts of chemicals

With the help of the balanced equation for the reaction, decide on the concentrations of solutions to use bearing in mind that these will normally be in the range you have met before while learning titration techniques and procedures. Choose concentrations that will lead to a titre of around 25 cm³.

Calculate the masses of any solids that need to be weighed out and dissolved to prepare solutions with your chosen concentration.

Decide whether any of the solutions provided need to be diluted. Give details to show how you would carry out the dilution.

If analysing solids or solutions such as medicines, cleaning agents, health care products, water samples, and so on, you may need to do some small-scale preliminary tests to get some idea of how much of a solid to dissolve or how much of a solution to dilute.

Choice of equipment

You need to show that you appreciate the accuracy and limitations of the equipment you use. You can do this by indicating which of the measurements of mass or volume need to be made as precisely as possible and which can be approximate.

For example, when using the iodine–thiosulfate method of estimating oxidising agents the potassium iodide is added in excess and so can be measured approximately, while the volume of thiosulfate solution is critical and is measured with a burette.

See Chapter 12 for details about the uncertainties in the use of volumetric glassware.

Detecting end-points

In redox and complex-forming titrations you are unlikely to have a choice of indicator or method of detecting the end-point, but in acid–base titrations you may be expected to make the choice.

In any acid–base titration there is a sudden change of pH at the end-point. For a titration of a strong acid with a strong base the pH jumps from around pH 3 to pH 10 at the end-point. Most common indicators change colour sharply in this range.

For a strong acid with a weak base the pH jump is not so great and is from around pH 3 to pH 8; while for a weak acid with strong base the jump is from about pH 6 to pH 10.

Indicators change colour over a range of around two pH units (Table 3.2). The chosen indicator has to complete its colour change within the range of values spanned at the end-point.

Table 3.2 Some common indicators and the pH range over which they change colour

Indicator	Colour change low pH–high pH	pH range over which colour change occurs
methyl orange	red–yellow	3.2–4.2
methyl red	yellow–red	4.8–6.0
bromothymol blue	yellow–blue	6.0–7.6
phenolphthalein	colourless–red	8.2–10.0

Hazards and safety precautions

Follow the guidelines in Chapter 11, bearing in mind that the hazards of acids, alkalis and other reagents vary with concentration.

Implementing

You must develop your titration technique through practice and take pride in your ability to get consistent results. You should aim to achieve two titres that agree to within 0.1 cm³.

It is very important to present your results in a neat table showing both burette readings for all the titrations you carry out including the rough

Note

When using a pipette:

- use a safety filler
- check cleanliness
- rinse with the solution to be measured
- line up the bottom of the meniscus with the graduation at eye level
- allow to drain under gravity
- touch the tip of the pipette with the surface of the solution in the flask.

titration. Always show that you have read the burette to the nearest half-scale division; that is to the nearest 0.05 cm³. Above the table of the results state clearly what was in the burette, what was in the flask and how the end-point was detected.

Example

In the flask: 20.0 cm³ of a solution of partially decomposed sodium carbonate crystals (2.696 g in 250 cm³ solution)
In the burette: 0.10 mol dm⁻³ hydrochloric acid

Titration number	1 (rough)	2	3
Final burette reading/cm³	22.00	23.00	22.15
Initial burette reading/cm³	1.00	2.35	1.60
Titre/cm³	21.00	20.65	20.55

The mean titre from the second and third values = 20.60 cm³

Analysing and drawing conclusions

In order to analyse the results of a titration successfully you must have a clear understanding of the design and purpose of the experiment. From the titration itself you can either work out the concentration, in moles per litre, of one of the solutions, or you can work out the ratio of the amounts of the two main reactants.

You can combine this information with what else you know about the solutions and the reactants as illustrated by the examples in this chapter.

You must select from your titration results the values for the titre which are sufficiently precise to allow you to calculate an accurate mean value.

You must always relate the calculations to the balanced equation for the titration reaction.

You must follow the guidance in Chapter 12 to estimate the overall percentage uncertainty in the calculated result and quote the final answer to the number of significant figures consistent with the precision of the experiment.

Evaluation

Follow the general advice given in Chapter 1. Where possible compare your final answer to published values, formulae or equations if known.

Discuss possible reasons for any differences between your results and those expected.

Suggest possible improvements to the design of the procedure, or the apparatus used, which might improve the results.

Sometimes you may be asked to evaluate an account of a titration carried out by someone else. If so, check through all aspects of the procedure and analysis to identify any points where the reported analysis did not correspond to best practice.

Test Yourself

3.1 The results in the example on page 28 come from a titration to investigate a sample of sodium carbonate crystals, $Na_2CO_3.10H_2O$, which had lost part of its water of crystallisation on exposure to air. Determine the percentage of water lost from the crystals from the titration results.

3.2 A 4.10 g sample of phosphonic acid, H_3PO_3, was dissolved in water and the volume of solution made up to $100\ cm^3$. $20.0\ cm^3$ of the acid solution was required to neutralise $25.0\ cm^3$ of a solution of sodium hydroxide containing 8.00 g of the alkali in $250\ cm^3$ solution. What was the equation for the reaction between the acid and the alkali?

3.3 A 1.576 g sample of ethanedioic acid crystals, $(COOH)_2.nH_2O$, was dissolved in water and made up to $250\ cm^3$. In a titration, $25.0\ cm^3$ of the acid solution reacted exactly with $15.6\ cm^3$ of $0.160\ mol\ dm^{-3}$ sodium hydroxide solution. What is the value of n in the formula for the acid?

3.4 Sodium ethanedioate, $Na_2C_2O_4$, is a primary standard that can be used to standardise solutions of potassium manganate(VII). Under acidic conditions, $KMnO_4$ oxidises ethanedioate ions on heating to carbon dioxide. In a titration it was found that $28.85\ cm^3$ of a solution of $KMnO_4$ oxidised $25.00\ cm^3$ of a solution containing $7.445\ g\ dm^{-3}$ of sodium ethanedioate. What was the concentration of the potassium manganate(VII) solution?

3.5 A $25.0\ cm^3$ sample of a solution containing iron(II) and iron(III) ions was titrated with potassium manganate(VII) in the presence of acid. The titre was $18.00\ cm^3$ with a $0.0200\ mol\ dm^{-3}$ solution of $KMnO_4(aq)$. A second $25.0\ cm^3$ sample of the original solution was reduced with zinc and acid. After filtering off the zinc, the solution was titrated with the same solution of $KMnO_4(aq)$. This time the mean titre was $22.50\ cm^3$. Calculate the concentrations of the iron(II) and iron(III) ions in the original solution.

3.6 Copper(II) ions oxidise iodide ions to iodine. A pale, off-white precipitate of a copper compound forms at the same time. A 3.405 g sample of copper(II) sulfate-5-water was dissolved in water and made up to $250\ cm^3$. Excess potassium iodide was added to a $25.0\ cm^3$ sample of the copper(II) sulfate solution. In a titration, $18.0\ cm^3$ of a solution of $0.0760\ mol\ dm^{-3}$ sodium thiosulfate was required to react with the iodine formed. What is the oxidation state of the copper in the precipitated copper compound formed during the reaction of copper(II) ions with iodide ions?

3.7 Several $100\ cm^3$ samples of hard water were titrated with $0.010\ mol\ dm^{-3}$ edta. The mean titre was $29.20\ cm^3$. What was the concentration of calcium ions in the hard water? What was the degree of hardness, expressed in terms of grams of calcium carbonate per litre?

CHAPTER FOUR

Enthalpy changes

Principles

Measuring energy changes is important because it helps chemists to explain the stability of compounds and to predict the likely direction of chemical change.

Enthalpy changes

The enthalpy change, ΔH, for any process is the exchange of energy between a reaction mixture (the system) and its surroundings when the change takes place at constant pressure.

- If the reaction is exothermic the system loses energy to its surroundings and ΔH is negative.

- If the reaction is endothermic the system gains energy from its surroundings and ΔH is positive.

When measuring enthalpy changes the energy given out (or taken in) by a reaction heats (or cools) a measured volume of water in an insulated container.

The standard enthalpy changes, $\Delta H^{\ominus}$, in databases are calculated for carefully specified conditions and for precise amounts of chemicals (in moles). The standard conditions for thermochemistry are 298 K and 1 bar pressure (100 kPa).

Hess's law

It is not possible to measure all enthalpy changes directly. Many values are determined indirectly using Hess's law. This law states that the enthalpy change for a reaction is the same whether it takes place in one step or in a series of steps. The overall enthalpy change is determined by the starting materials and the products and not by the method of converting one to the other.

> **Hint**
>
> In thermochemistry temperatures are always measured on the Kelvin scale.
>
> $0\,^{\circ}\text{C} = 273\,\text{K}$

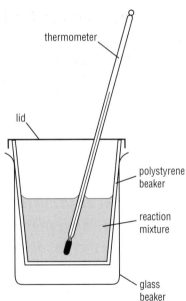

thermometer

lid

polystyrene beaker

reaction mixture

glass beaker

Figure 4.1 *A simple calorimeter for measuring enthalpy changes in solution*

Methods

Apparatus

A calorimeter is an apparatus for measuring the energy change during a chemical change. The name means 'calorie-measurer' and dates back to the days when chemists measured energy changes in calories instead of joules.

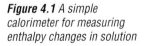

It is possible to measure enthalpy changes in solution with a fair degree of accuracy using a simple calorimeter consisting of an expanded polystyrene cup with a lid and a thermometer. Using two cups, one inside the other, improves the insulation (Figure 4.1).

Expanded polystyrene is an excellent insulator and has a negligible specific heat capacity. If the reaction is exothermic, then the energy from the reaction heats up the solution in the cup. By measuring the temperature rise it is possible to calculate the energy change.

The temperature rises are relatively small, so for accurate work the thermometer should be graduated to tenths of a degree. Alternatively an accurately calibrated digital thermometer can be used.

Calculation

The calculation for reactions involving dilute solutions is based on these assumptions:

- all the energy from an exothermic reaction transfers to the solution and none heats the cup or is lost to the surroundings

- the density of the solution and its specific heat capacity can be taken to be the same as for pure water.

The energy change $= mc\Delta T$

In this relationship:

- m is the mass of water in the plastic cup which can be calculated from the total volume of solution after mixing, assuming that the density is $1\,\mathrm{g\,cm^{-3}}$

- c is the specific heat capacity of water which is $4.18\,\mathrm{J\,g^{-1}\,K^{-1}}$

- ΔT is the measured temperature change

Then enthalpy change for the particular reaction is the energy change for the amounts (in moles) shown in the chemical equation:

$$\text{Enthalpy change} = \frac{\text{calculated energy change}}{\text{amount (in moles) of the limiting reagent}}$$

> **Hint**
>
> The extent of reaction is usually controlled by a limiting reagent. Other reactants are in excess. It is only necessary to determine the amount (in moles) of the limiting reagent.

4.3 Enthalpy change of a redox reaction

Purpose

The aim here is to measure the enthalpy change for the oxidation of zinc metal by copper(II) ions.

Method

Figure 4.2 outlines the procedure. In this instance the zinc is in excess. The copper(II) ions are the limiting reactant. The amount of copper(II) is determined by the volume and concentration of the solution of copper(II) sulfate.

Check safety before carrying out any practical procedure.

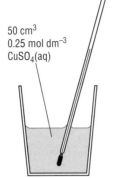
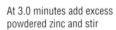

50 cm³
0.25 mol dm⁻³
$CuSO_4(aq)$

excess powdered
zinc

Figure 4.2 Outline
procedure for measuring
the enthalpy change of a
reaction

Measure the temperature
every 30 s for 2.5 minutes

At 3.0 minutes add excess
powdered zinc and stir

Continue stirring and record
the temperature every 30 s
for a further 6 minutes

Results

The graph in Figure 4.3 shows the temperatures recorded before and
after adding the powdered zinc. The solution was close to room
temperature before mixing and its temperature barely changed.
After mixing it took a little while for the temperature to reach a peak
and then it began to cool.

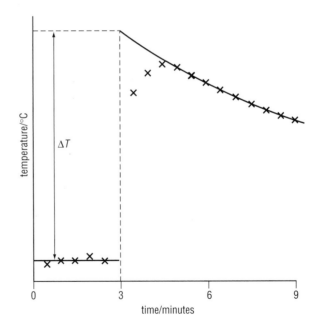

Figure 4.3 Plot to show
temperature changes
before and after adding the
zinc powder to the copper
sulfate solution

Extrapolating the two lines on the graph to 3 minutes makes it possible to
estimate ΔT, which is the maximum temperature rise assuming that the
metal all reacted at once and there was no loss of energy from the
calorimeter to the surroundings. Here $\Delta T = 12.5$ K.

Calculation

$$Zn(s) + Cu^{2+}(aq) \rightarrow Zn^{2+}(aq) + Cu(s)$$

The solution gets hotter during the reaction. This is an exothermic reaction. So ΔH is negative.

The energy given out $= 50.0\,g \times 4.18\,J\,g^{-1}\,K^{-1} \times 12.5\,K = 2612\,J$

The zinc is in excess. The amount of copper(II) limits the reaction.

The amount of copper(II) reacting $= \dfrac{50}{1000}\,dm^3 \times 0.25\,mol\,dm^{-3} = 0.0125\,mol$

The energy change for the amounts in the equation $= \dfrac{2612\,J}{0.0125\,mol}$

$$= 209000\,J\,mol^{-1}$$

$$= 209\,kJ\,mol^{-1}$$

For $Zn(s) + Cu^{2+}(aq) \rightarrow Zn^{2+}(aq) + Cu(s)$, $\Delta H = -209\,kJ\,mol^{-1}$

4.4 Applying Hess's law

Purpose

The aim is to determine the enthalpy change for the hydration of magnesium sulfate to give crystals of the hydrated salt:

$$MgSO_4(s) + 7H_2O(l) \rightarrow MgSO_4.7H_2O(s)$$

It is not possible to measure this enthalpy change directly because of the difficulty of controlling the temperature and measuring the temperatures of solids. By applying Hess's law it is possible to determine the required enthalpy change at room temperature.

The required enthalpy change, ΔH_1 can be calculated by finding the values of ΔH_2 and ΔH_3 (Figure 4.4).

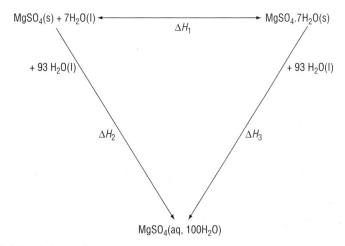

Figure 4.4 Hess's law cycle

Check safety before
carrying out any
practical procedure.

Method in outline

Figure 4.5 shows the procedure for determining ΔH_2. Exactly the same
procedure can be used to determine ΔH_3 using hydrated magnesium sulfate
in place of the anhydrous salt and 93 mol water in place of 100 mol. For both
parts of the experiment the solid has to be weighed accurately because it is
the limiting reagent.

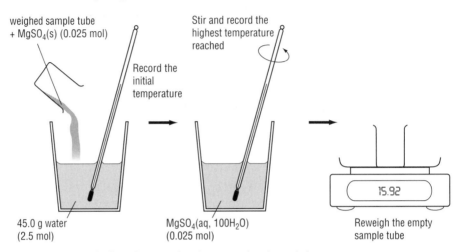

weighed sample tube
+ MgSO$_4$(s) (0.025 mol)

Record the
initial
temperature

Stir and record the
highest temperature
reached

45.0 g water
(2.5 mol)

MgSO$_4$(aq, 100H$_2$O)
(0.025 mol)

15.92

Reweigh the empty
sample tube

Figure 4.5 *Outline of a procedure for measuring the enthalpy
change when anhydrous magnesium sulfate reacts with and
dissolves in 100 mol water*

Results

Table 4.1 *Results for anhydrous and hydrated salt*

Results for anhydrous salt		Results for hydrated salt	
mass of sample tube + MgSO$_4$(s)	12.91 g	mass of sample tube + MgSO$_4$.7H$_2$O(s)	19.58 g
mass of empty sample tube	15.92 g	mass of empty sample tube	13.42 g
mass of MgSO$_4$(s) added	3.01 g	mass of MgSO$_4$.7H$_2$O(s) added	6.16 g
mass of polystyrene cup + water	47.10 g	mass of polystyrene cup + water	44.21 g
mass of empty cup	2.10 g	mass of empty cup	2.36 g
mass of water	45.00 g	mass of water	41.85 g
temperature of the solution after reaction	35.4 °C	temperature of the solution after reaction	23.4 °C
starting temperature of the water	24.1 °C	starting temperature of the acid	24.8 °C
temperature change	+11.3 K	temperature change	−1.4 K

Calculation

Calculating ΔH_2

Energy given out by the reaction $= 45.0\,\text{g} \times 4.18\,\text{J}\,\text{g}^{-1}\,\text{K}^{-1} \times 11.3\,\text{K}$
$$= 2130\,\text{J} = 2.13\,\text{kJ}$$

Amount of $MgSO_4(s)$ added $= \dfrac{3.01\,\text{g}}{120\,\text{g}\,\text{mol}^{-1}} = 0.0250\,\text{mol}$

Enthalpy change for the exothermic reaction, ΔH_2
$$= \frac{-2.13\,\text{kJ}}{0.0250\,\text{mol}} = -85.0\,\text{kJ}\,\text{mol}^{-1}$$

Calculating ΔH_3

Energy taken in by the reaction $= 41.85\,\text{g} \times 4.18\,\text{J}\,\text{g}^{-1}\,\text{K}^{-1} \times 1.4\,\text{K}$
$$= 245\,\text{J} = 0.245\,\text{kJ}$$

Amount of $MgSO_4.7H_2O(s)$ added $= \dfrac{6.16\,\text{g}}{246\,\text{g}\,\text{mol}^{-1}} = 0.0250\,\text{mol}$

Enthalpy change for the exothermic reaction, ΔH_3
$$= +\frac{0.245\,\text{kJ}}{0.0250\,\text{mol}} = +9.8\,\text{kJ}\,\text{mol}^{-1}$$

Calculating ΔH_1

According to Hess's law: $\Delta H_1 + \Delta H_3 = \Delta H_2$

Hence: $\Delta H_1 = \Delta H_2 - \Delta H_3$.

$$\Delta H_1 = (-85.0\,\text{kJ}\,\text{mol}^{-1}) - (+9.8\,\text{kJ}\,\text{mol}^{-1}) = -94.8\,\text{kJ}\,\text{mol}^{-1}$$

4.5 Practical skills

Experiments to measure energy changes in solution are often used as planning exercises. They can also be used to assess the other practical skills. The following section suggest key points to attend to when you are being assessed on practical activities to measure enthalpy changes.

Planning

The apparatus

A polystyrene cup is fine for this kind of practical work. Note that you will use a lid to limit exchange of energy with the surroundings. You can improve the insulation by using two cups, one inside the other.

Remember to specify an accurate thermometer capable of measuring changes to the nearest tenth of a degree.

You will need a burette, or pipette with safety filler, to measure the volumes of solutions accurately.

The amounts of chemicals

Always write the balanced equation for the reaction.

Always show your plan to your teacher before starting any practical work.

Compare in your mind the volume of a typical plastic cup with the volumes of standard laboratory beakers. This will help you to specify suitable volumes of solutions. Choose concentrations in the range you have experienced in previous practical work.

Decide which reactant is to be the limiting reagent and then use the equation to calculate the amount of the other chemical needed to ensure that it is in excess (see Chapter 2).

The procedure

Always write out a step-by-step procedure which someone else could follow without further guidance. Describe the procedure of measuring a series of temperatures before and after mixing the chemicals to make allowance for energy transfer between the surroundings and the mixture in the cup.

One or more labelled diagrams can be very helpful in making your intentions clear. When illustrating the measurement of temperature always make sure that the bulb of the thermometer (or the whole sensor) is shown as immersed in the liquid.

Hazards and safety precautions

Follow the guidelines in Chapter 11, bearing in mind that the hazards of acids, bases and salts in solution vary with concentration. Always mention eye protection if it is needed because of the hazards. Specify the use of a pipette filler when using a pipette.

Test Yourself

4.1 Magnesium reacts with oxygen: $Mg(s) + \frac{1}{2}O_2(g) \rightarrow MgO(s)$.

Your task is to plan an experiment to determine the enthalpy change for this reaction by an indirect method. Assume that you have access to: magnesium ribbon, magnesium oxide powder and $1.0 \, mol \, dm^{-3}$ hydrochloric acid together with the usual apparatus in an advanced level laboratory.

Include in your plan:

- the quantities of chemicals to be used

- step-by-step instructions that someone else could follow

- a statement of precautions taken to ensure that the results are as accurate as possible

- a clear explanation of how you would use your measurements to work out the enthalpy for the reaction shown above

- a note of the hazards and the safety precautions to be taken to reduce the risks from the hazards.

Implementing

Measure the temperatures particularly carefully, trying to read your thermometer at least to the nearest half-scale division.

Analysing and drawing conclusions

Recording results

Record your results in a table. Make sure that you have a clear heading for each column or row. Always include the units with measurements.

Displaying the results

Plot a graph of the series of temperature measurements before and after mixing. Choose the axes carefully and label them clearly. Extrapolate the two lines on each graph to arrive at an estimate of the true temperature change for the amounts of chemicals you have mixed.

The calculation

Write the balanced equation for the change and use it to identify the limiting reagent which has controlled the extent of change.

State the assumptions you are making in carrying out the calculation.

Set out the calculation clearly, step-by-step, with a few words at the start of each line to show what it means. Include the units with all measurements.

Calculate the energy change for the amounts used in the experiment and then scale up to the amounts in the balanced equation.

Check that you present your final answer with an appropriate number of significant figures.

Remember that ΔH refers to the amounts in the equation. Make sure that you show the correct sign for ΔH.

Test Yourself

4.2 Calculate the enthalpy change of neutralisation of hydrochloric acid by sodium hydroxide, given that on mixing 200.0 cm^3 of 0.40 mol dm^{-3} hydrochloric acid with 200.0 cm^3 0.40 mol dm^{-3} sodium hydroxide solution the temperature after mixing was 26.6 °C when the temperature of both solutions before mixing was 23.9 °C.

4.3 Calculate the enthalpy change of solution of potassium chloride given that the temperature of 25.0 cm^3 water fell from 24.3 °C to 22.1 °C on adding 1.00 g of the salt and stirring until it dissolved.

Evaluation

Estimating uncertainty

Follow the guidelines in Chapter 12 to estimate the measurement uncertainty associated with your experiment. Identify the main sources of error in both the measurements and the procedure.

Commenting on the apparatus and procedure

Look critically at the method you followed and suggest ways of minimising errors and increasing reliability. Suggest improvements that could be made to the experimental procedures.

Comment on the assumptions made in calculating the results and state whether or not you think that the assumptions are justified.

Example

The Hess's law experiment described on pages 33–35 gives a value for $\Delta H_1 = -94.8 \text{ kJ mol}^{-1}$.

The accepted value of $\Delta H_1 = -104 \text{ kJ mol}^{-1}$.

So the overall percentage error $= \dfrac{(-104.0) - (-94.8)}{104.0} = -8.8\%$

The measurable apparatus uncertainties arise from measuring temperatures to the nearest 0.1 K and masses of the salts to 0.1 g.

An estimate of the uncertainties in the determination of ΔH_2 are 0.2 K in the measurement of a temperature difference of 11.3 K (percentage uncertainty $\approx$ 2%) and 0.1 g in the measurement of 3.01 g (percentage uncertainty $\approx$ 3%). Total uncertainty = 5%.

An estimate of the uncertainties in the determination of ΔH_3 are 0.2 K in the measurement of a temperature difference of 1.4 K (percentage uncertainty $\approx$ 14%) and 0.1 g in the measurement of 6.16 g (percentage uncertainty $\approx$ 2%). Total uncertainty = 16%.

Thus the combined estimated uncertainty can more than account for the observed discrepancy. However these estimates are pessimistic as explained on page 104.

Note that in the procedure described, no attempt was made to allow for energy exchange with the surroundings by plotting a series of temperature readings before and after each change, as illustrated in Figure 4.3. Following a procedure of this kind has the potential to decrease the uncertainty in the measurement of the temperatures which contributes the largest percentage uncertainty to the overall result.

Another possible source of uncertainty to consider here is that the $MgSO_4(s)$ may not have been completely anhydrous because it tends to pick up moisture from the air.

CHAPTER FIVE

Reaction kinetics

5.1 Principles

Studying rates of reaction is important because it helps chemists to control reactions both in laboratories and on a large scale in industry. Kinetic studies also help chemists to work out the mechanisms of reactions.

Rate equations

Chemists have found that they can summarise their findings about reaction rates with the help of rate equations. In general for a reaction

$$aA + bB \rightarrow products$$

the rate equation takes the form:

$$rate = k[A]^n[B]^m$$

The powers n and m are the reaction orders. The overall order is (n + m).

The rate constant, k, is only constant at a particular temperature.

In advanced chemistry courses there are two main purposes of practical work exploring kinetics. One is to determine reaction orders by investigating how reaction rates vary with concentration. The other is to determine activation energies by investigating how rate constants vary with temperature.

Activation energies

Reaction rates vary with temperature because the value of the rate constant, k, varies with temperature. The Arrhenius equation describes the relationship between k and temperature, T.

$$\ln k = \text{constant} - \frac{E_A}{RT}$$

E_A is the activation energy for the reaction, T the temperature on the Kelvin scale and R the gas constant. A plot of $\ln k$ against $\frac{1}{T}$ gives a straight-line graph. The gradient of the line is $-\frac{E_A}{R}$.

5.2 Methods

Concentration–time graphs

A concentration–time graph shows either how the concentration of a product increases with time, or how the concentration of a reactant falls with time. The gradient (slope) of a concentration–time graph at any point measures the rate of a reaction at that time.

> **Note**
>
> It is impossible to predict the order of a reaction with respect to each reactant from the balanced equation. Orders can only be found by experiment.

> **Note**
>
> The general equation for a straight line takes the form:
>
> $$y = mx + c$$
>
> where m is the gradient of the line and c a constant.
>
> With the usual axes, the gradient is positive if the line slopes up from left to right and negative if it slopes down.

One practical method to monitor the changes in concentration over time is to remove samples from the reaction mixture and then to stop the reaction in the samples by rapidly cooling or by changing the pH.

Another method can be used if the reaction produces a gas. Figure 5.1 shows the apparatus that can be used to study the rate of reaction between marble chips and dilute hydrochloric acid. Marble and acid are added to the flask. The reaction is allowed to proceed for a short while to saturate the solution with carbon dioxide, then the bung is put in place and timing starts.

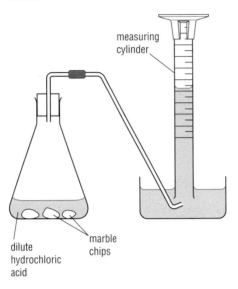

measuring cylinder

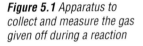

Figure 5.1 *Apparatus to collect and measure the gas given off during a reaction*

dilute hydrochloric acid

marble chips

Typically the marble chips are in excess. The reaction slows down and stops as the hydrochloric acid gets less concentrated and then is finally used up all together.

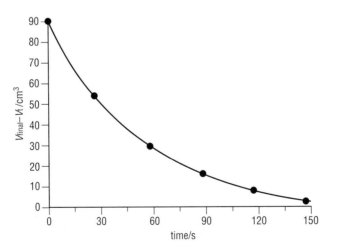

Figure 5.2
Concentration–time graph for the reaction of marble chips with acid

The total volume of gas collected when the reaction stops = V_{final}. This final volume is proportional to the hydrochloric acid concentration when the bung was put in place and timing started. The greater the concentration of acid at the start the more gas will eventually form.

At any time t after the start of timing the volume of gas collected is V_t. This varies with the amount of acid that has reacted by that time. V_t increases as t increases while the acid concentration falls. So $(V_{final} - V_t)$ is proportional to the concentration of hydrochloric acid at time t.

So plotting $(V_{final} - V_t)$ against t gives a concentration–time graph (Figure 5.2) for the experiment.

Initial-rate methods

Determining the initial rate for a kinetics experiment is often important. One way to find the initial rate is to draw a tangent at the start of a concentration–time graph and use it to calculate the gradient at time zero.

This can be laborious and there is a useful short cut which simplifies the design of rate experiments.

Figure 5.3 shows two plots of the amount of product formed with time. Line 1 shows the formation of a product under one set of conditions. An amount of product x forms in time t_1. Line 2 shows the formation of the same product under a different set of conditions. The same amount of product x forms in the longer time t_2.

The average rate of formation of product on line $1 = \dfrac{x}{t_1}$

The average rate of formation of product on line $2 = \dfrac{x}{t_2}$

If x is kept the same, it follows that the average rate near the start $\propto \dfrac{1}{t}$

This means that it is possible to arrive at a measure of the initial rate of a reaction by measuring how long the reaction takes to produce a small, fixed amount of product, or use up a small, fixed amount of reactant.

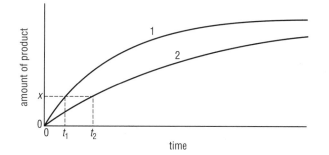

Figure 5.3 Two graphs showing the formation of a product with time, for the same reaction under different conditions

This technique can be used to study the rate of reaction of magnesium with an acid. The acid has to be in significant excess so that its concentration does not change during the reaction. The procedure is to measure the time for a small, fixed mass of magnesium ribbon to dissolve completely. The experiment is repeated with the same mass of metal and volume of acid but with varying acid concentrations.

The approach can also be used to study the reaction of sodium thiosulfate with acid. This reaction slowly produces a precipitate of sulfur. A simple but effective procedure is to measure the time taken for enough sulfur to form to obscure a cross underneath the flask containing the reaction mixture. It is reasonable to assume that the same amount of sulfur is needed to hide the cross each time (Figure 5.4).

Figure 5.4 *An arrangement for measuring the initial rate of the reaction of acid with sodium thiosulfate*

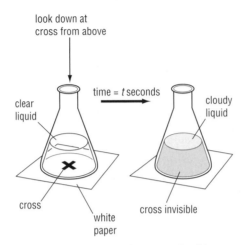

A variant on the initial-rate method is to use a 'clock reaction', so called because the reaction is set up to produce a sudden colour change after a certain time when it has produced a fixed amount of one reactant.

The reaction of peroxodisulfate(VI) ions with iodide ions can be set up as a clock reaction:

$$S_2O_8^{2-}(aq) + 2I^-(aq) \rightarrow 2SO_4^{2-}(aq) + I_2(aq)$$

A small, known amount of sodium thiosulfate ions is added to the reaction mixture which also contains starch indicator. At first the thiosulfate reacts with any iodine, I_2, as soon as it is formed turning it back to iodide ions (see page 23), so there is no colour change. At the instant when all the thiosulfate has been used up, free iodine is produced and this immediately gives a deep blue-black colour with the starch. If t is the time for the blue colour to appear after mixing the chemicals, then once again $\frac{1}{t}$ is a measure of the initial rate of reaction.

Check safety before carrying out any practical procedure.

5.3 Investigating orders of reaction

Purpose

This example shows how the 'clock' method for determining initial rates can be used to find the reaction orders in the rate equation for the oxidation of bromide ions by bromate(V) ions under acid conditions:

$$5Br^-(aq) + BrO_3^-(aq) + 6H^+(aq) \rightarrow 3Br_2(aq) + 3H_2O(l)$$

The problem is to find the values of p, q and r in the rate equation:

$$Rate = k[Br^-(aq)]^p[BrO_3^-(aq)]^q[H^+(aq)]^r$$

Method

Setting up the clock

The reaction is set up for the 'clock' method by adding two other chemicals to the reaction mixture: a small, measured amount of phenol and methyl orange indicator.

At the start, any bromine formed reacts very rapidly with the phenol present. Free bromine does not appear in the solution until all the phenol is used up. Immediately there is any free bromine it bleaches the indicator so the solution turns from pink to colourless.

All-the-reactants-kept-constant-except-one procedure

When studying this reaction it would be impossible to interpret the data if all the concentrations of the three reactants were allowed to vary together. To make it possible to analyse the results, the procedure is to carry out three distinct series of experiments. In each series, the initial concentration of one of the reactants is varied systematically while the concentrations of the other reactants are kept constant.

Finding the order with respect to bromide ions

Figure 5.5 shows how a series of six runs with six different mixtures can give a set of values for the initial rate of reaction for six different bromide ion concentrations, while all the other concentrations are kept the same.

Figure 5.5 Mixtures for exploring the effect of bromide ion concentration on the initial rate of reaction

Run	Volume of 0.01 mol dm^{-3} KBr/cm^3	Volume of water/cm^3
1	10.0	0.0
2	8.0	2.0
3	6.0	4.0
4	5.0	5.0
5	4.0	6.0
6	3.0	7.0

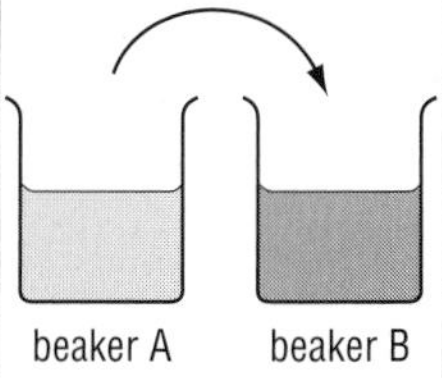

beaker A   beaker B

Run	Volume of 0.005 mol dm^{-3} KBrO$_3$/cm^3	Volume of solution of acid and methyl orange/cm^3	Volume of 0.00010 mol dm^{-3} phenol/cm^3
1	10.0	15.0	5.0
2	10.0	15.0	5.0
3	10.0	15.0	5.0
4	10.0	15.0	5.0
5	10.0	15.0	5.0
6	10.0	15.0	5.0

Note that in beaker B the volumes and concentrations are the same for all six runs. In beaker A, however, the volume of the potassium bromide ion solution varies but the total volume is kept the same by adding enough water to ensure that the volume of the reaction mixture is the same for each run.

For each run, the contents of beakers A and B are mixed and timing started. Timing stops when the pink colour of the indicator disappears.

After each run the temperature of the mixture is recorded to ensure that it stays the same throughout the series of experiments.

Finding the orders with respect to bromate(v) and hydrogen ions

Another set of experiments investigates the effect of systematically changing the bromate(v) ion concentration. In this case beaker A contains a range of mixtures of potassium bromate(V) and water, while beaker B always contains the same volumes of the solutions of potassium bromide, acid with methyl orange and phenol.

A final set of experiments investigates the effect of changing the concentration of hydrogen ions. In these runs beaker A contains a series of mixtures of acid and water while beaker B contains, each time, the same volumes of solutions of potassium bromate(V), potassium bromide and phenol solution.

Results and analysis

The table of results shows the time for the indicator colour to disappear for the series of experiments to investigate how the initial rate varies with the bromide ion concentration. As explained on page 41, $\frac{1}{t}$ is a measure of the initial rate. The temperature of the solutions was $21.5 \pm 0.5\,°C$.

> **Note**
>
> Multiplying the calculated values of $\frac{1}{t}$ in the results table by a constant factor makes the numbers easier to plot without changing the gradient.

Table 5.1 Time taken for indicator colour to disappear and the initial rate

Volume of Br⁻(aq)/cm³ as in Figure 5.5	10.0	8.0	6.0	5.0	4.0	3.0
t/s	22.5	26.0	34.0	37.5	48.0	65.0
$\frac{1}{t} \times 100$/s⁻¹	4.44	3.85	2.95	2.67	2.08	1.54

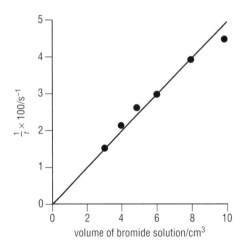

Figure 5.6 Plot to explore the order of reaction with respect to bromide ions

Since the concentrations of bromate(V) and hydrogen ions were constant, the rate equation simplifies to:

$$\text{Rate} = \text{constant} \times [\text{Br}^-(\text{aq})]^p$$

In this investigation $\frac{1}{t}$ measures the rate, while the concentration of bromide ion is proportional to the volume of KBr(aq) in the mixture which has a constant total volume. So the rate equation becomes:

$$\frac{1}{t} = \text{constant} \times [\text{volume of KBr(aq)}]^p$$

In the first instance it is worth trying out a plot of rate $(\frac{1}{t})$ against bromide ion concentration (measured by the volume of aqueous KBr). The plot turns out to be a straight line as shown in Figure 5.6, the value of p is 1 and the reaction is first order with respect to bromide ions.

A similar analysis of the results from the second set of experiments produces a similar straight-line graph showing that the reaction is also first order with respect to bromate(v) ions.

A plot of rate against hydrogen ion concentration for the third series of results does not produce a straight line. The reaction is not first order with respect to hydrogen ions.

In this set of experiments the rate equation simplifies to:

$$\text{Rate} = \text{constant} \times [\text{H}^+(\text{aq})]^r$$

Taking into account the form of the results, this becomes:

$$\frac{1}{t} = \text{constant} \times [\text{volume of acid}]^r$$

Taking logs of both sides of this equation gives;

$$\log(\tfrac{1}{t}) = \log(\text{constant}) + r\log[\text{volume of acid}]$$

Plotting $\log(\frac{1}{t})$ against log(volume of acid) gives a straight-line graph with gradient r.

> ### Note
>
> Logarithms are defined in such a way that:
>
> $\log ab = \log a + \log b$
>
> and that:
>
> $\log a^n = n\log a$
>
> This applies both to logarithms to base 10 (log) and natural logarithms to base e (ln).

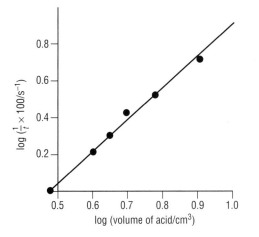

Figure 5.7 *A logarithmic plot to determine the order of the reaction with respect to hydrogen ions*

The gradient of the graph in Figure 5.7 is 1.8. It may be that the order with respect to hydrogen ions is 2; however there are reactions with orders that are not whole numbers.

The complete set of results suggests that the rate equation is:

$$\text{Rate} = k[\text{Br}^-(\text{aq})][\text{BrO}_3^-(\text{aq})][\text{H}^+(\text{aq})]^2$$

The overall order of reaction is 4.

5.4   Measuring activation energies

Purpose

This example shows how the 'clock' method for determining initial rates can be used to find the activation for the oxidation of iodide ions by peroxodisulfate(VI) ions:

$$S_2O_8^{2-}(aq) + 2I^-(aq) \rightarrow 2SO_4^{2-}(aq) + I_2(aq)$$

The investigation can be extended to study the catalytic effect of d-block element ions on the reaction. Effective catalysts provide an alternative pathway for the change with a lower activation energy.

Check safety before carrying out any practical procedure.

Method

In this series of experiments the concentrations of the reactants are kept constant while the temperature of the reaction mixture is varied systematically over an appropriate range of values.

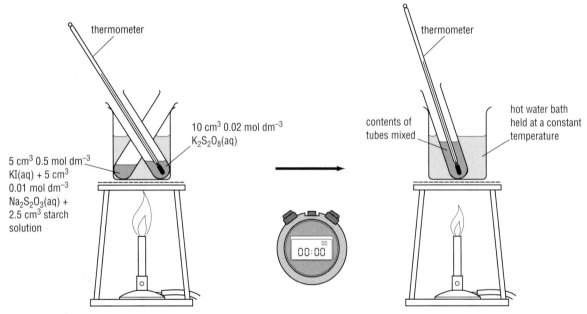

thermometer

10 cm³ 0.02 mol dm⁻³ K₂S₂O₈(aq)

contents of tubes mixed

thermometer

hot water bath held at a constant temperature

5 cm³ 0.5 mol dm⁻³ KI(aq) + 5 cm³ 0.01 mol dm⁻³ Na₂S₂O₃(aq) + 2.5 cm³ starch solution

00:00

Figure 5.8 *Outline of a procedure for investigating how the rate of a reaction varies with temperature*

The reaction mixture includes starch solution. Adding a small, measured amount of sodium thiosulfate as well means that at the start any iodine formed is immediately reduced back to iodide ions. Once all the thiosulfate has been used up, the free iodine formed suddenly gives a blue-black colour with the starch (Figure 5.8).

Results

Table 5.2 shows a typical set of results from a series of runs with temperatures in the range 30 °C and 55 °C. The table includes calculated values for $\ln(\frac{1}{t})$ and $\frac{1000}{T}$ K^{-1} corresponding to the variables in the Arrhenius equation (see page 39).

Table 5.2

Temperature, T/K	303	309	312	318	324
Time, t, for the blue colour to appear/s	204	138	115	75	55
$\ln(\frac{1}{t})$		−5.32	−4.93	−4.75	−4.32	−4.01
$\frac{1000}{T}$/K^{-1}	3.30	3.24	3.21	3.14	3.09

Note

When using the Arrhenius equation, temperatures must be in kelvin. Multiplying the values of $\frac{1}{T}$ by a constant factor makes them easier to plot without affecting the gradient of the resulting graph.

Analysis

A plot of $\ln(\frac{1}{t})$ against $\frac{1}{T}$ gives a straight line which has a negative gradient (Figure 5.9).

The magnitude of the slope works out to be -6.36×10^3 K.

So $-\dfrac{E_A}{R} = -6.36 \times 10^3$ K

Hence $-E_A = -6.36 \times 10^3$ K $\times 8.31$ J K^{-1} mol^{-1}

The activation energy, $E_A = 52.9$ kJ mol^{-1}

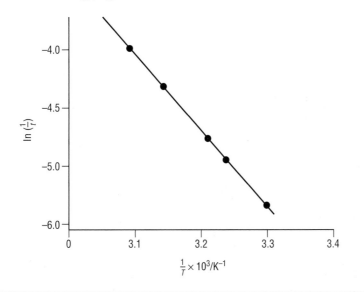

Figure 5.9 Arrhenius plot to determine the activation energy for a reaction

---

5.5   Practical skills

Planning

Rates of reaction vary with concentration, particle size of any solids, temperature and the presence of catalysts. With so many variables it is important to have a plan that varies one factor at a time while keeping the other variables constant.

One approach to kinetics experiments is to concentrate on the initial rate. On mixing the reagents at the start all the concentrations are known. Clock reactions are a variant on the initial-rate method (see pages 41–45).

Always show your plan to your teacher before starting any practical work.

Checklist

When planning rate experiments make sure that you:
★ explain the design of your experiment and state which factors you will vary and which variables you will control
★ show how you have used the trial tests and/or the equation for the reaction to decide on the concentrations and quantities to use
★ list the apparatus you need and draw a labelled diagram to show how it will be used
★ give step-by-step instructions for carrying out the experiment
★ indicate the measurements you plan to take
★ explain how you will analyse your results, showing what graphs you will plot and why
★ identify any hazards and state how you will reduce the risks from them.

An alternative approach is to follow the change in concentration of one reactant until it is nearly used up with all the other reactants present in such large excess that their concentrations do not change to a significant extent. You can then plot a concentration–time graph for the one reactant with varying concentration, and plot tangents to the curve to measure the gradient, and hence the rate, at a succession of concentrations.

Sometimes it is easier to follow the formation of a product, for example, by collecting and measuring the volume V of a gas, and then use the fact that, as explained on pages 40–1, $(V_{final} - V_t)$ is proportional to the concentration of one of the reactants at time t.

Always carry out some preliminary tests to find out (by experiment) a suitable range of concentrations or temperatures that will give you a reasonable range of results in times that you can measure with sufficient accuracy.

Write out the balanced equation and use it to estimate suitable quantities to use. If, for example, you are planning to study a reaction that produces a gas you will need to ensure that you will not produce more gas than you can sensibly collect and measure.

The balanced equation also allows you to check the extent to which concentrations will vary during an experimental run. If studying the time for a measured amount of metal to dissolve in an acid, you need to check that your acid is in sufficiently large excess for its concentration not to vary before all the metal is used up.

You should show which measurements are critically important in determining the final result. You should decide when to use a burette or pipette to measure volumes of solutions and when to use a measuring cylinder.

Implementing

You have to be well organised and practically competent to get good results from kinetics experiments. You have to be able to measure volumes and times accurately. If you are following a reaction over a period of time you have to be alert to take a succession of readings with appropriate precision at regular intervals.

Some techniques are demanding, such as removing samples from a reaction mixture with a pipette, stopping the reaction in some way and then titrating the sample to measure the concentration of a reactant.

It is important to measure the temperature of a reaction mixture to ensure that it does not vary substantially. If the reaction is exothermic this may affect an experiment which you are trying to carry out at constant temperature.

Analysing and drawing conclusions

You will usually need to set out your results in a neat table with extra rows or columns to calculate quantities such as $\frac{1}{t}$ or $(V_{\text{final}} - V_t)$. Always label the columns and rows clearly and give the units for physical quantities.

In any kinetics experiment you are likely to have to draw and interpret one or more graphs. These can include plots of:

- concentration against time – the gradient at any point is the rate at that time

- rate against concentration – a straight line through the origin for a first order reaction, a horizontal line for a zero order reaction

- log(rate) against log(concentration) – with a gradient that equals the order with respect to the particular reactant

- ln (rate) against $\frac{1}{T}$ – the gradient is used to calculate the activation energy (see page 47).

After plotting appropriate graphs from your results you should show how they lead to your conclusions about the orders of reaction or the activation energy, giving any working in full. Always remember the units for physical quantities.

Your analysis should include estimates of the main sources of measurement uncertainty (see Chapter 12) together with a value for the overall uncertainty.

You should also follow the guidance in Chapter 1. In some cases your analysis can attempt to discuss the significance of the results. If you have found the orders of reaction with respect to the various reactants you should attempt to discuss the implications of the overall rate equation with respect to possible mechanisms for the reaction.

If you have measured the activation energy for a reaction with and without a catalyst, you can speculate on reasons why the catalyst proved effective and provided a reaction pathway with a lower activation energy.

Test Yourself

5.1 The rate of reaction of magnesium with hydrochloric acid was investigated by adding 5 cm lengths of magnesium ribbon to excess dilute acid. The hydrogen gas was collected in a gas syringe. The table shows the times taken to collect 10 cm^3 hydrogen using a range of concentrations of acid.

continued ➤

Test Yourself *continued*

Concentration of HCl(aq)/mol dm^{-3}	0.2	0.4	0.6	0.8	1.0
Time to collect 10 cm^3 hydrogen/s	60	30	20	15	12

By plotting a suitable graph, find out what the results suggest about the order of the reaction with respect to hydrogen ions.

5.2 The table shows the results of a series of experiments to determine the activation energy for the oxidation of iodide ions by peroxodisulfate(VI) ions in the presence of iron(III) ions (see section 5.4).

Temperature, T/K	288	292.5	299	308	315
Time, t, for the blue colour to appear/s	10.0	7.0	5.0	3.5	2.5

(a) Analyse the results and plot a graph to find a value for the activation energy in the presence of iron(III) ions.

(b) How does the value of the activation energy compare with the value for the reaction in the absence of iron(III) ions? (see page 47).

(c) Suggest an explanation for the effect of adding iron(III) ions.

Evaluation

Commenting on the reliability of data
You should review your findings and identify any anomalous results. You should then discuss your estimates of measurement uncertainty and comment on whether the degree of uncertainty casts significant doubts on your conclusions.

If you have measured the activation energy for a reaction you should work out the overall percentage uncertainty in the result.

Comparing outcomes with expectations
When studying some reactions you may be able to find information about the expected order of reaction or the accepted value for the activation energy. If so, you should discuss how your findings compare with the results reported elsewhere.

Identifying possible improvements
You should take a critical look at the design of the experiment and the practical methods involved. Aim to suggest improvements to minimise errors and increase reliability.

Possible investigations

There are only a limited number of reactions that lend themselves to kinetic experiments with the equipment and time available in advanced chemistry courses. Below are some pointers to help you get started with your planning for an investigation in this field. Other possibilities include the examples featured earlier in the chapter which are not discussed further here.

Reaction of magnesium with acid

The reaction of magnesium with acid is a popular reaction for investigations. As it is a reaction that is also studied in more elementary courses, it is important to make sure that you approach your investigation of this change in a way that connects with the theory you have learned during your advanced chemistry course.

There is a discussion of approaches to this investigation in the Practical Investigations section of the Re:act website (see www.chemistry-react.org/go/Topic/Default_4.html). Particularly helpful is the article that you can download from the same page of that website. This reports on a series of studies of metal–acid reactions with a discussion of the results. The paper will help you to identify a worthwhile focus for your investigation. Also, when it comes to evaluation, you will be able to compare your findings with those in the article and discuss similarities and differences.

Decomposition of hydrogen peroxide

A solution of hydrogen peroxide, H_2O_2, does not decompose at a measurable rate if stored in a clean container. The compound is unstable and decomposes rapidly in the presence of some inorganic catalysts such as manganese(IV) oxide. The enzyme called catalase is a particularly effective catalyst for this decomposition reaction. The rate of reaction can be followed by collecting and measuring the oxygen produced. If you plot the volume of oxygen against time, you can then measure the initial rate from the gradient of a tangent to the curve drawn at time $t = 0$ s. You will need to carry out preliminary experiments to find a suitable concentration range for the hydrogen peroxide.

In your investigation you might:

- find the order of the reaction with respect to hydrogen peroxide

- study the effect of varying the concentration of a finely dispersed source of catalase (such as active yeast or potato)

- identify a range of catalysts for the reaction and see how the presence of a catalyst affects the activation energy for the reaction.

Oxidation of iodide ions by hydrogen peroxide

The oxidation of iodide ions by hydrogen peroxide is catalysed by hydrogen ions. This can be set up as a clock reaction by adding a small, measured amount of sodium thiosulfate to the reaction mixture with a little starch solution (see page 42).

Consult reference books, or use trial and error, to find suitable concentrations and volumes of the solutions. In container A mix your hydrogen peroxide solution with dilute sulfuric acid. In container B mix potassium iodide solution, distilled water with the fixed volume of sodium thiosulfate solution and starch solution. Mix the contents of the two containers to start the reaction and measure the time for the reaction to turn blue.

Vary the iodide concentration by adjusting the volumes of potassium iodide solution and water in the second container while keeping the total volume constant so that the other concentrations do not vary.

In a similar way set up the contents of containers A and B to investigate the effect of the hydrogen peroxide concentration on the rate.

If you try to investigate the effect of hydrogen ions on the rate you will find the results less easy to interpret. It is an interesting challenge to try to work out how the rate varies with the hydrogen ion concentration, and why.

Reaction of iodine with propanone
The substitution reaction of iodine with propanone is catalysed by acid. There are several ways of following the course of the reaction that make it possible to determine the order of reaction with respect to iodine, propanone and hydrogen ions.

The simplest approach is to estimate the initial rate by measuring the time for the iodine colour to disappear under different conditions with propanone solution, dilute hydrochloric acid and water in a flask to which a measured volume of iodine is quickly added to start the reaction.

A more systematic approach uses a titration method. Timing starts on mixing the contents of two flasks: one containing a mixture of solutions of propanone, dilute acid and water; the other containing the solution of iodine. After mixing, samples are withdrawn with a pipette at regular intervals and run into a solution of sodium hydrogencarbonate which neutralises the acid catalyst and stops the reaction. The iodine concentration at that time can then be measured by titration with sodium thiosulfate.

A series of experiments makes it possible to determine the initial rate under different conditions as the initial concentration of one reactant, say propanone, is varied while the concentrations of the other chemicals are kept the same. Hence the order of reaction with respect to propanone can be determined. Then the whole procedure is repeated first for iodine and then for hydrogen ions.

A third method of following the reaction is to use a colorimeter. This requires some skill but can give very good results. The great advantage is that the reaction can be followed by measuring the absorption of light by the iodine in the solution without having to interfere with the reaction mixture by withdrawing samples for analysis. Only small volumes of solutions are needed when the reaction mixture is contained in a colorimeter tube.

Hydrolysis of methyl ethanoate
Methyl ethanoate hydrolyses fast enough at room temperature in the presence of an acid catalyst for it to be possible to study the kinetics of the

reaction and determine the order of the reaction with respect to the ester. The reaction produces ethanoic acid and so can be followed by titrating samples withdrawn from the reaction mixture with a standard solution of sodium hydroxide. Before the titration it is necessary to stop the reaction; this can be done by cooling the sample in iced water.

CHAPTER SIX
Equilibria

Principles

The equilibrium law

The equilibrium law is a quantitative law for predicting the amounts of reactants and products present when a reversible reaction comes to a state of equilibrium.

For the reaction: $aA + bB \rightleftharpoons cC + dD$

$$K_c = \frac{[C]^c[D]^d}{[A]^a[B]^b}$$

K_c is the equilibrium constant when the concentrations of reactants and products are measured in moles per litre.

K_c is a constant at a given temperature but its value varies as the temperature changes. This accounts for the effect of temperature changes on the position of equilibrium.

This law is widely applicable and is particularly helpful in making sense of the behaviour of weak acids and bases, indicators and buffer solutions. The law applies to all reversible reactions at equilibrium, but the scope for practical investigations of the law is relatively limited within the time and facilities available in advanced chemistry courses.

Weak acids and bases

Weak acids are only very slightly ionised in aqueous solution. The equilibrium law applies. For a weak acid HA:

$$HA(aq) + H_2O(l) \rightleftharpoons H_3O^+(aq) + A^-(aq)$$

$$K_c = \frac{[H_3O^+(aq)][A^-(aq)]}{[HA(aq)][H_2O(l)]}$$

In dilute solutions the concentration of water is effectively constant, so that the expression can be rewritten in this form where K_a is the acid dissociation constant:

$$K_a = \frac{[H_3O^+(aq)][A^-(aq)]}{[HA(aq)]}$$

Indicators (represented as HIn) are weak acids which change colour when they lose protons. In other words the colour of HIn(aq) is distinct from the colour of In$^-$(aq). Indicators are added a drop or two at a time so they have to adjust to the hydrogen ion concentration (and so the pH) set by the other molecules and ions in solution. The colour of an indicator at a particular pH is determined by the value K_a for the indicator (often written as K_{In}).

Buffer solutions are solutions present in large enough amounts to dictate the pH. A typical buffer solution is a mixture of a weak acid, HA(aq), and one of its salts supplying the conjugate base, A⁻(aq). The ratio of the concentrations of the base and the acid controls the hydrogen ion concentration (and hence the pH) as shown by the equilibrium law in this arrangement:

$$[H_3O^+(aq)] = \frac{K_a\,[HA(aq)]}{[A^-(aq)]}$$

6.2 Measuring K_c

The strategy for determining the value of an equilibrium constant involves three main steps:

- **Step 1**: Mix measured quantities of reactants and/or products. Then allow the mixture to come to equilibrium under steady conditions.

- **Step 2**: Analyse the mixture to find the equilibrium concentration of one of the chemicals at equilibrium.

- **Step 3**: Use the equation for the reaction and the information from steps 1 and 2 to work out the values for the equilibrium concentrations of all the atoms, molecules or ions. Then substitute these values into the expression for K_c.

The challenge when investigating reactions at equilibrium is to find ways to measure equilibrium concentrations without upsetting the equilibrium. Many methods of analysis use up the chemical being analysed. Analytical methods of this kind are generally unsuitable because, as Le Chatelier's principle shows, the position of equilibrium shifts whenever one of the reactants or products is removed from the equilibrium mixture.

There are two main ways to measure equilibrium concentrations. One way is to find a way of 'freezing' the reaction and thus slowing down the rate so much that it is possible to measure one of the equilibrium concentrations by titration. The most obvious way to slow down the rate and 'freeze' the equilibrium is by cooling. Other possibilities are to dilute the equilibrium mixture or to remove a catalyst.

The second way of measuring equilibrium concentrations is to use an instrument that responds to a property of the solution that varies with concentration. Well established methods for doing this include measuring the pH with a pH meter or measuring the intensity of a colour with a colorimeter.

Note

Le Chatelier's principle is a quick way of predicting the consequences of the equilibrium law.

6.3 Ester formation

Purpose

The equilibrium between ethanoic acid, ethanol, ethyl ethanoate and water is one of the few reaction systems (other than acid–base equilibria) that lends itself to study in an advanced chemistry course:

$$CH_3COOH(l) + C_2H_5OH(l) \rightleftharpoons CH_3COOC_2H_5(l) + H_2O(l)$$

The reaction is very, very slow at room temperature in the absence of a catalyst. In the presence of an acid catalyst the reaction mixture comes to equilibrium in about 48 hours.

Diluting the equilibrium mixture means that the reaction is slow enough to find the equilibrium concentration of ethanoic acid by titration without the position of equilibrium shifting perceptibly in the time taken for the titration.

Method

Step 1: Mix measured quantities of chemicals and allow the mixture to come to equilibrium.

Precisely measured quantities of the chemicals are added to sample tubes. The masses of the components of the mixture can be found by weighing or by measuring their volume and using densities to find the masses. The sample tubes are tightly stoppered to avoid loss by evaporation and set aside at constant temperature for 48 hours.

Some of the tubes might at first contain just ethanol, ethanoic acid and hydrochloric acid. Others might start with only ethyl ethanoate, water and hydrochloric acid. Working in this way shows that it is possible to reach equilibrium from either side of the equation.

Step 2: Analyse the mixture to find the equilibrium concentration of the acid.

Each equilibrium mixture is transferred quantitatively to a flask and diluted with water. Titration with a standard solution of sodium hydroxide determines the total amount of acid in the sample at equilibrium: both hydrochloric acid and ethanoic acid.

Step 3: Use the equation for the reaction and the information from steps 1 and 2 to work out the values for all the equilibrium concentrations.

Some of the sodium hydroxide used in the titration reacts with the hydrochloric acid. Since the amount of HCl(aq) does not change, it is possible to work out how much of the titre was used to neutralise it, knowing how much HCl(aq) was added at the start. The remainder of the alkali added during the titration reacts with ethanoic acid. Hence the amount of ethanoic acid at equilibrium can be calculated. The other equilibrium concentrations can be found given the starting amounts of chemicals present and the equation for the reaction.

Results and analysis

The approach to the calculation can be illustrated from the results for a sample which initially contained the following:

- ethyl ethanoate, 3.64 g (0.0413 mol)

- water, 0.99 g

- 5 cm³ of 2.0 mol dm⁻³ HCl(aq) containing 0.010 mol HCl (mass 0.365 g).

The mass of the added hydrochloric acid = 5.17 g

So this contained 5.17 g − 0.365 g $\qquad$ = 4.81 g water

So the total mass of water at the start $\qquad$ = 0.99 g + 4.81 g = 5.80 g

$\qquad\qquad\qquad\qquad$ = 0.322 mol H_2O

Titration of the equilibrium mixture found that 39.20 cm^3 of 1.00 mol dm^{-3} sodium hydroxide neutralised the total acid present. Both HCl and CH_3COOH react 1 : 1 with NaOH so this shows that the total amount of acid at equilibrium was:

$$\frac{39.2}{1000} \text{ dm}^3 \times 1.00 \text{ mol dm}^{-3} = 0.0392 \text{ mol}$$

So, taking away the amount of hydrochloric acid added at the start, this shows that the amount of ethanoic acid at equilibrium is:

$$0.0392 \text{ mol} − 0.010 \text{ mol} = 0.0292 \text{ mol}$$

	$CH_3COOH(l)$ +	$C_2H_5OH(l)$ ⇌	$CH_3COOC_2H_5(l)$	+ $H_2O(l)$
Initial amount/mol	0.00	0.00	0.0413	0.322
Measured amount at equilibrium/mol	0.0292			
Other equilibrium amounts found from the starting amounts and the equation/mol		0.0292	0.0413 − 0.0292 = 0.0121	0.322 − 0.0292 = 0.293

Each of these equilibrium amounts should be converted to equilibrium concentrations, in mol dm^{-3}, by dividing by the total volume, V. However here it is not necessary to know the volume because the volume cancels in the expression for the equilibrium concentration since there are two concentration terms on the top line and two on the bottom. This also means that in this example K_c has no units:

$$K_c = \frac{[CH_3COOC_2H_5(l)][H_2O(l)]}{[CH_3COOH(l)][C_2H_5OH(l)]} = \frac{(0.0121/V)(0.293/V)}{(0.0292/V)(0.0292/V)} = 4.16$$

Repeating the determination of K_c with a range of starting mixtures of chemicals makes it possible to demonstrate that it is indeed a constant.

6.4 Measuring K_a for a weak acid

Purpose

The aim is to measure the value of K_a for a weak acid such as ethanoic acid:

$$CH_3COOH(aq) + H_2O(l) ⇌ H_3O^+(aq) + CH_3COO^-(aq)$$

The forward and back reactions are fast and so the system comes to equilibrium immediately. This means that it is not possible to determine the equilibrium concentration of hydrogen ions by titration with alkali. As soon as hydrogen ions are neutralised more ethanoic acid ionises. This goes on until all the ethanoic acid is converted to a salt. So a titration of dilute ethanoic acid with an alkali such as sodium hydroxide measures the total amount of acid present in the solution:

$$CH_3COOH(aq) + NaOH(aq) \rightarrow CH_3COO^-Na^+(aq) + H_2O(l)$$

It is possible to estimate the value of K_a for a weak acid by using a meter to measure the pH of a dilute solution of the acid with a known concentration. However, this does not give accurate results since the pH is easily affected by other dissolved substances including, for example, carbon dioxide from the air.

Method

A good method to determine K_a for a weak acid is to measure the pH of the solution in the flask during a titration of the acid with a dilute solution of a strong alkali, such as sodium hydroxide.

Step 1: Mix measured quantities of chemicals and allow the mixture to come to equilibrium.

Check safety before carrying out any practical procedure.

Measured volumes of sodium hydroxide are run into a measured volume of the weak acid solution in the flask. The added alkali neutralises some of the acid and turns it into its sodium salt. After each addition of alkali there is a new equilibrium mixture in the flask.

Step 2: Analyse the mixture to find the equilibrium concentration of the acid.

All that is necessary is to measure the pH after each addition of alkali:

$$pH = -\log[H_3O^+(aq)]$$

Step 3: Use the equation for the reaction and the equilibrium law to find the value of the equilibrium constant.

As shown on page 55 the expression for K_a can take the form:

$$[H_3O^+(aq)] = \frac{K_a[HA(aq)]}{[A^-(aq)]}$$

Taking logarithms gives:

$$\log[H_3O^+(aq)] = \log K_a + \log\frac{[HA(aq)]}{A^-(aq)]}$$

which is equivalent to:

$$pH = pK_a + \log\frac{[A^-(aq)]}{[HA(aq)]}$$

When $[A^-(aq)] = [HA(aq)]$ this becomes $pH = pK_a + \log 1$

which means that $pH = pK_a$ when half the acid has been neutralised and converted to its salt so that $[A^-(aq)] = [HA(aq)]$.

This makes it possible to read a value for pK_a directly from a graph showing the pH plotted against the titre.

Results

Figure 6.1 shows the results of plotting pH against titre for a titration of 25.0 cm^3 of a roughly 0.1 mol dm^{-3} solution of chloroethanoic acid with 0.10 mol dm^{-3} sodium hydroxide solution.

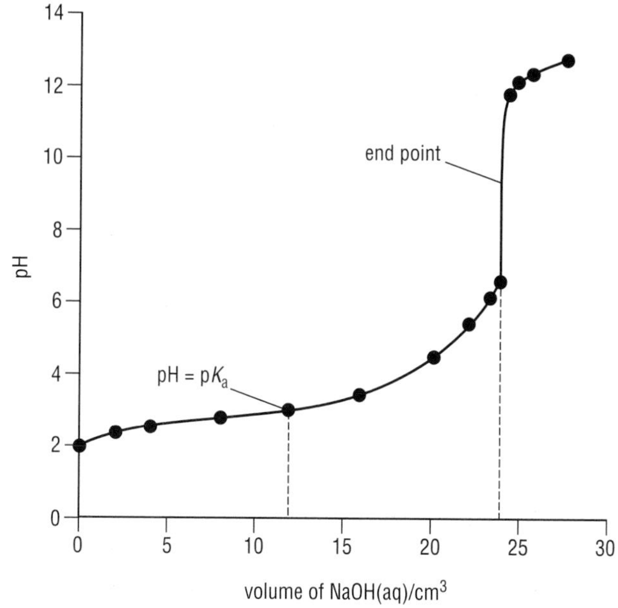

Figure 6.1 *Plot of pH against titre for a titration of chloroethanoic acid with sodium hydroxide*

> **Note**
>
> In this method for determining K_a it is not necessary to know the concentrations of the acid or the alkali accurately so long as the pH values during the titration are measured precisely (to the nearest 0.1 unit) and the value of the end-point is determined accurately.

The end-point of the titration is at 24.0 cm^3. This is the titre corresponding to the point halfway up the steeply rising part of the curve.

This means that half the acid was neutralised and converted to its salt after adding 12.0 cm^3 of the alkali. At this point on the graph the pH = 2.9. So:

$$pK_a = -\log K_a = 2.9$$

$$K_a = 10^{-2.9} = 1.3 \times 10^{-3}\ \text{mol dm}^{-3}$$

> **Note**
>
> $K_a = 10^{-pK_a}$
>
> The units of K_a are mol dm^{-3}.

6.5 Measuring K_{In} for an indicator

Purpose

The aim is to determine the dissociation constant, K_{In}, for an indicator. The indicator, HIn, is a weak acid.

$$HIn(aq) + H_2O(l) \rightleftharpoons H_3O^+(aq) + In^-(aq)$$

The logarithmic form of the equilibrium law can be used to determine the value of K_{In}:

$$pH = pK_{In} + \log\frac{[In^-(aq)]}{[HIn(aq)]}$$

Check safety before
carrying out any
practical procedure.

Method

The method here works for an indicator for which HIn and In⁻ each have a
distinct colour. A suitable indicator is bromophenol blue which is yellow at
low pH values and blue at high pH values.

The approach is to find the ratio of the concentrations of the two forms of
the indicator at a known pH. This can be done using a home made colour
matching system as shown in Figure 6.2.

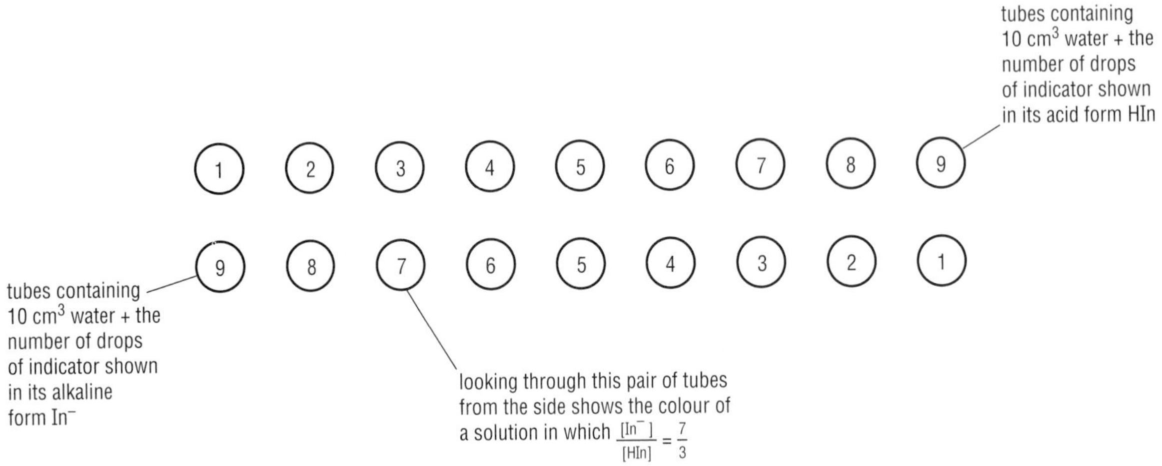

tubes containing
10 cm³ water + the
number of drops
of indicator shown
in its acid form HIn

(1) (2) (3) (4) (5) (6) (7) (8) (9)

(9) (8) (7) (6) (5) (4) (3) (2) (1)

tubes containing
10 cm³ water + the
number of drops
of indicator shown
in its alkaline
form In⁻

looking through this pair of tubes
from the side shows the colour of
a solution in which $\frac{[\text{In}^-]}{[\text{HIn}]} = \frac{7}{3}$

*Figure 6.2 Circles
represent matched test
tubes each containing
10 cm³ water with the
indicated number of drops
of the indicator in either its
HIn or In⁻ form*

Step 1: Mix measured quantities of chemicals and allow the mixture to
come to equilibrium.

The equilibrium is set up by pouring 10 cm³ of a buffer solution at pH 3.7
into a test tube (the same volume as the colour matching tubes). Then ten
drops of the indicator are added (the same number of drops as the total for
each pair of colour matching tubes).

Step 2: Analyse the mixture to find the equilibrium concentration of the
acid.

After mixing, the procedure is to compare the colour of the indicator in the
buffer solution with the colour seen when looking at a white surface
through both of each pair of test tubes, as illustrated in Figure 6.2. The aim
is to find the best match.

Step 3: Use the equation for the reaction and the information from steps 1
and 2 to work out the values for all the equilibrium concentrations.

The best match for the indicator in the buffer solution should be with the
colour seen when looking through the pairs of tubes with three drops of
indicator in the alkaline tubes and seven drops in the acid tubes.

Calculation

The best colour match gives:

$$\frac{[\text{In}^-(\text{aq})]}{[\text{HIn}(\text{aq})]} = \frac{3}{7}$$

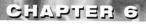
The buffer solution chosen has a pH of 3.7:

$$3.7 = pK_{In} + \log \frac{3}{7}$$

$$pK_{In} = 3.7 - \log \frac{3}{7} = 3.7 + 0.37 = 4.1$$

So for bromophenol blue:

$$K_{In} = K_a = 10^{-4.1} = 7.9 \times 10^{-5}\,mol\,dm^{-3}$$

6.6 Practical skills

Planning

Planning investigations of equilibrium systems does not feature largely in advanced chemistry courses because of the limited number of experiments that are feasible in practice.

If you are asked to make plans in this area you should follow the three-step procedure illustrated by the examples in this chapter.

What you may be asked to do is to suggest variants or extensions of established methods.

Always show your plan to your teacher before starting any practical work.

Test Yourself

6.1 (a) Show that the set of pairs of tubes in figure 6.2 can be used as a pH meter covering the range 3.0–5.0 once the value of pK_{In} is known.

(b) Describe the steps you would follow to use the pairs of tubes in Figure 6.2 to measure the dissociation constant K_a for a weak acid such as methanoic acid.

Implementing

Very careful measurement of all masses and volumes is essential for success in experiments of this kind. It is crucial to follow precisely the given procedure.

It is also important to record all measurements that you make, such as balance readings and burette readings. Set the values out clearly in a table and then show the masses and volumes calculated from the measurements.

Analysing and drawing conclusions

You need to have a grasp of the theory underlying the design of the experiment when working out the results of experiments to investigate equilibria.

Keep track of the logic of calculations by setting them out neatly, step-by-step. Always include the units for physical quantities.

Estimate the measurement uncertainty in the values that critically affect your answer (see Chapter 12). Work out a value for the total measurement uncertainty and quote your final answer with an appropriate number of significant figures.

Check the form of the equilibrium law expression if your answer is a value for an equilibrium constant to make sure that you give correct units for K_c.

Test Yourself

6.2 Titration of the equilibrium in another sample tube from the experiment of Section 6.2 required 41.30 cm^3 of 1.00 mol dm^{-3} NaOH(aq). Calculate a value for K_c given that, at the start, the tube contained 4.51 g ethyl ethanoate with 5.0 cm^3 of 2.0 mol dm^{-3} HCl(aq) but no added water other than the water in the dilute acid.

6.3 In a variant of the procedure described in Section 6.4, 5.0 cm^3 of $0.020 \text{ mol dm}^{-3}$ benzenecarboxylic acid was mixed with 5.0 cm^3 of $0.020 \text{ mol dm}^{-3}$ sodium benzenecarboxylate. On adding ten drops of bromophenol blue to the mixture, the colour matched the colour seen on looking through the pair of tubes with six drops of indicator in the alkaline tube and four drops in the acid tubes. Calculate K_a for benzenecarboxylic acid given that pK_a for the indicator is 4.1.

Evaluation

You should evaluate your experiments in this field following the advice on page 5 in Chapter 1.

Data books quote values for K_c, K_a or K_{In} for most of the systems you are likely to investigate. This means that it is possible to compare your results with the generally accepted value. You can discuss whether any discrepancy between the two values can be explained by your estimates of the total measurement uncertainty.

CHAPTER SEVEN

Inorganic reactions

7.1 | Principles

Part of the art of a chemist is to have a 'feel' for the way in which chemicals behave and to recognise characteristic patterns of behaviour. This helps you to know what to look for when making observations and helps you to spot any unexpected changes.

Inorganic tests are based on chemical reactions that produce colour changes, gases and precipitates. There are four main types of inorganic reaction, which you need to know about when interpreting observations while testing inorganic compounds.

Inorganic reactions used in tests

Ionic precipitation reactions
The simplest examples of precipitation reactions involve mixing two solutions of soluble compounds that are ionised in solution. The positive ions from one compound combine with the negative ions of the other compound to form an insoluble compound which comes out of solution as a precipitate.

This type of reaction is used to test for negative ions (anions). Adding a soluble barium salt (nitrate or chloride) to a solution of a soluble sulfate, for example, gives rise to a white precipitate of insoluble barium sulfate.

Precipitation reactions can also help to identify positive ions (cations). Adding a dilute solution of sodium hydroxide to a solution of a soluble metal salt produces a precipitate if the metal hydroxide is insoluble in water. Insoluble metal hydroxides may also precipitate with ammonia solution.

Acid-base reactions
Acids and alkalis are commonly used in chemical tests. Dilute hydrochloric acid is a convenient strong acid. Sodium hydroxide solution is often chosen as a strong base.

A carbonate is a salt of a weak unstable acid, carbonic acid. Adding dilute hydrochloric acid to a carbonate adds protons to the carbonate ions, CO_3^{2-}, turning them into carbonic acid molecules, H_2CO_3, which immediately decompose into carbon dioxide and water. Testing with limewater can then identify the gas given off, confirming that the compound tested is a carbonate. Sulfite, sulfides and nitrites can all be identified in a similar way.

Some of the insoluble metal hydroxides are amphoteric. If so, they dissolve in excess sodium hydroxide solution.

Many of the test reagents mentioned in this chapter are hazardous. Before carrying out any tests, check that you are aware of the hazards and are taking suitable precautions to reduce the risks. Never work unsupervised.

Redox reactions

Common oxidising agents used in inorganic tests include chlorine, bromine, and acidic solutions of the manganate(VII) ion or the dichromate(VI) ion.

Some reagents change colour when oxidised. This makes them useful for detecting oxidising agents. In particular a colourless solution of iodide ions turns to a yellow-brown colour when oxidised. This can be a very sensitive test if starch is present because starch gives an intense blue-black colour with iodine. This is the basis of the use of starch-iodide paper to test for chlorine and other oxidising gases.

Common inorganic reducing agents are metals (in the presence of acid or alkali), sulfur dioxide and iron(II) ions. A test for nitrate ions is based on the use of aluminium in alkali to reduce nitrate ions to ammonia.

Some reagents change colour when reduced. In particular dichromate(VI) ions in acid change from yellow to green. This is the basis of a test for sulfur dioxide. The colour change when dichromate(VI) ions are reduced is just one of many examples of transition metal ions changing colour when oxidised or reduced.

Complex-forming reactions

The formation of a complex ion can bring about two visible changes that can be useful in analysis. There may be a colour change or an insoluble compound may dissolve. Sometimes both types of change happen at the same time. For example, a pale blue precipitate of copper(II) hydroxide dissolves in excess ammonia solution and the solution turns deep blue because the copper(II) ions form a stable complex ion with ammonia molecules.

Intensely coloured transition metal complexes can be useful as tests for specific metal ions. Iron(III) ions, for example, form a very deep red complex with thiocyanate ions, SCN^-.

The common test for water is based on the formation of a complex ion. Anhydrous cobalt(II) chloride is blue. It turns pink in the presence of water as the cobalt ions form a complex with water molecules.

Note

Some transition elements show distinctive colours in their different oxidation states. An example is vanadium, V:

- V(V): yellow
- V(IV): blue
- V(III): green
- V(II): violet.

Note that the unknown substances you are asked to test may be hazardous or undergo hazardous reactions. Always wear eye protection. Assume the highest level of hazard. Remember that the nature of the hazard may vary with the concentration of solutions.

7.2 Tests and observations

Preliminary tests

Preliminary observations and tests provide a general introduction to the characteristics of a compound. They can include:

- the state and appearance of the compound at room temperature
- the solubility of the compound in cold and hot water and the pH of any solution that forms
- the effect of heating the compound.

Solubility in water

Chemists find it useful to apply a rough classification of solubility based on what they see on shaking a small amount of the solid with water in a test tube:

- **very soluble**, e.g. potassium nitrate; plenty of the solid dissolves quickly

- **soluble**, e.g. copper(II) sulfate; the crystals visibly dissolve to a significant extent and colour the solution if the compound is coloured

- **slightly soluble**, e.g. calcium hydroxide; the solid does not appear to dissolve but testing with indicator shows that enough has dissolved to affect the pH of the solution

- **insoluble**, e.g. aluminium oxide; where there is no sign that the solid dissolves and the pH of the water remains unchanged.

Familiarity with the patterns of solubility of common inorganic compounds helps with the interpretation of observations during test tube experiments (see Table 7.1).

Table 7.1 *Solubility of acids, bases and salts*

Type of compound	Soluble in water	Insoluble in water
acids	all common acids are soluble in water	
bases	soluble bases are alkalis and include the hydroxides of group 1 metals, including sodium and potassium, and the carbonates of sodium and potassium. Lithium, magnesium, calcium and barium hydroxide are slightly soluble. Ammonia is a soluble base	all other metal oxides and hydroxide
salts	all nitrates all chlorides … all sulfates … all sodium, potassium and ammonium salts	 … except silver and lead chlorides … except barium sulfate, lead sulfate and calcium sulfate which is slightly soluble all other carbonates, chromates, sulfides and phosphates

Effect of heating

Heating a small sample of a solid inorganic compound can provide clues to the identity of the compound (Table 7.2). If a gas or vapour is given off it can be identified with the help of the tests shown in Table 7.3 on pages 66–67.

Bear in mind that gases given off may be hazardous.

Table 7.2 *Some observations on heating and possible inferences*

Observation on heating	Possible inferences
steamy vapour which turns blue cobalt chloride paper pink. The solid may turn to a solution	crystals contain water of crystallisation or the solid is a hydroxide which decomposes
colourless gas evolves which relights a glowing splint	oxygen from the nitrate of sodium or potassium
brown gas evolves and a glowing splint relights	nitrogen dioxide and oxygen from the decomposition of a nitrate
colourless gas given off which turns limewater cloudy white	carbon dioxide from carbonate or hydrogencarbonate
colourless acidic gas given off which turns paper soaked in potassium dichromate(VI) solution from orange to green	sulfur dioxide from a sulfite or sulfate
residue turns yellow when hot but then white again when cool	zinc oxide from the decomposition of another zinc compound such as the hydroxide or carbonate
sublimate forms on a cool part of the tube	likely to be an ammonium salt

Tests for gases

Gases can be identified by their colour, their odour, their effect on moist indicator paper and by the results of specific confirmatory tests (Table 7.3).

Table 7.3 *Some tests for gases*

Gas	Test	Observations
hydrogen	burning splint	burns with a 'pop'
oxygen	glowing splint	splint bursts into flame (relights)
carbon monoxide	burning splint	burns with a blue flame but does not explode
carbon dioxide	limewater (aqueous calcium hydroxide)	turns milky white as a precipitate forms
hydrogen halides	smell	pungent
	moist blue litmus paper	turns red
	ammonia vapour (from a drop of concentrated ammonia solution on a glass rod)	thick white smoke of particles of solid ammonium halide
	drop of silver nitrate on a glass rod	turns cloudy: white for HCl, cream if HBr and yellow if HI

Table 7.3 *Some tests for gases – continued*

Gas	Test	Observations
chlorine	colour	pale greenish-yellow
	smell	pungent, bleach-like
	moist blue litmus paper	turns red and then bleaches
	moist starch iodide paper	turns blue-black
bromine	colour	orange-brown
	moist blue litmus paper	turns red and then slowly bleaches
iodine	colour	violet
	moist starch (or starch iodide) paper	turns blue-black
sulfur dioxide	smell	pungent
	moist blue litmus paper	turns red
	paper soaked in potassium dichromate(VI) solution	turns from orange to green
hydrogen sulfide	smell	'bad eggs'
	burning splint	gas burns; yellow deposit of sulfur
	paper soaked in lead(II) ethanoate solution	turns brown-black
ammonia	smell	pungent and eye-watering
	moist red litmus paper	turns blue
nitrogen dioxide	colour	orange-brown
	moist blue litmus paper	turns red
water vapour	appearance	'steams' in the air
	anhydrous cobalt(II) chloride paper	turns from blue to pink

Tests for negative ions (anions)

Salts of weak acids, such as carbonates, sulfites, sulfides and nitrites, react with dilute hydrochloric acid, which is a strong acid. The stronger acid displaces the weaker acid which may then be identifiable itself or decompose to a recognisable gas.

Salts of some strong acids such as halide ions and sulfates can be identified by precipitation reactions. Salts of nitric acid can be identified by reducing the nitrate ion to ammonia.

Table 7.4 *Some tests for salts of weak and strong acids*

Test	Observations	Inferences
Test for carbonate, sulfide, sulfite or nitrite add dilute hydrochloric acid to the solid salt. Warm gently if there is no reaction at room temperature	gas which turns limewater cloudy white	carbon dioxide from a carbonate
	gas which smells foul and turns lead ethanoate paper black	hydrogen sulfide from a sulfide
	gas which is acidic, has a pungent smell and turns acid-dichromate paper from orange to green	sulfur dioxide from a sulfite
	colourless gas given off which turns brown where it meets the air	nitrogen oxide (NO) from a nitrite turning to nitrogen dioxide (NO_2)
Test for halide ions make a solution of the salt. Acidify with dilute nitric acid, then add silver nitrate solution. Test the solubility of any precipitate with ammonia solution	white precipitate soluble in dilute ammonia solution	precipitate of AgCl from a chloride
	cream precipitate soluble in concentrated ammonia solution	precipitate of AgBr from a bromide
	yellow precipitate insoluble in excess ammonia solution	precipitate of AgI from an iodide
Test for sulfate ions make a solution of the salt. Add a solution of barium nitrate or chloride. If a precipitate forms add dilute nitric acid	white precipitate which does not re-dissolve in acid	precipitate of $BaSO_4$ from a sulfate
Test for nitrate ions make a solution of the salt. Add sodium hydroxide solution and then a piece of aluminium foil or some Devarda alloy	alkaline gas evolves which turns litmus blue and has a pungent smell	ammonia from the reduction of a nitrate (or nitrite)

Tests for positive ions (cations)

Tests with aqueous alkalis

Aqueous sodium hydroxide and aqueous ammonia can help to identify metal ions in solution.

Adding sodium hydroxide solution produces a precipitate if the hydroxide of the metal is insoluble in water. The precipitate dissolves in excess of the alkali if the hydroxide is amphoteric.

Adding ammonia solution also precipitates insoluble hydroxides. These redissolve in excess of the reagent if the metal ions form stable complex ions with ammonia molecules.

Mixing an ammonium salt with sodium hydroxide solution produces free ammonia molecules because hydroxide ions are a stronger base than ammonia. Warming drives off the ammonia as a gas which can be detected by its smell and effect on moist red litmus paper.

Table 7.5 *Some tests for positive ions in solution*

Possible ion in solution	Observations on adding sodium hydroxide solution drop-by-drop and then in excess	Observations on adding ammonia solution drop-by-drop and then in excess
calcium, Ca^{2+}	white precipitate but only if the calcium ion concentration is high	no precipitate
magnesium, Mg^{2+}	white precipitate, insoluble in excess reagent	white precipitate, insoluble in excess reagent
barium, Ba^{2+}	no precipitate	no precipitate
aluminium, Al^{3+}	white precipitate which dissolves in excess reagent	white precipitate, insoluble in excess reagent
chromium(III), Cr^{3+}	green precipitate which dissolves in excess reagent to form a dark green solution	green precipitate, insoluble in excess reagent
manganese(II), Mn^{2+}	off-white precipitate, insoluble in excess reagent, precipitate turns brown on standing	off-white precipitate, insoluble in excess reagent
iron(II), Fe^{2+}	green precipitate, insoluble in excess reagent, which turns brown at the surface on standing	green precipitate, insoluble in excess reagent
iron(III), Fe^{3+}	browny-red precipitate insoluble in excess reagent	browny-red precipitate, insoluble in excess reagent
cobalt(II), Co^{2+}	blue precipitate from a red solution, insoluble in excess	blue precipitate from a red solution dissolving in excess to give a blue solution
nickel(II), Ni^{2+}	green precipitate, insoluble in excess reagent	green precipitate which dissolves in excess reagent to give a deep blue solution
copper(II), Cu^{2+}	pale blue precipitate, insoluble in excess reagent	pale blue precipitate dissolving in excess reagent to form a dark blue solution
zinc, Zn^{2+}	white precipitate which dissolves in excess reagent	white precipitate which dissolves in excess reagent
lead(II), Pb^{2+}	white precipitate which dissolves in excess reagent	white precipitate, insoluble in excess reagent
ammonium, NH_4^+	alkaline gas (ammonia) given off on heating	no visible change

Flame tests

Some metal ions do not give precipitates with alkalis. These ions can be identified by flame tests. Flame tests can also distinguish metal ions which behave in the same way with solutions of sodium hydroxide or ammonia.

Table 7.6 *Identifying metal ions with flame tests*

Metal ion	Flame colour
lithium	bright red
sodium	bright yellow
potassium	pale mauve
calcium	orange-red
strontium	scarlet
barium	apple green
copper(II)	green with flashes of blue

7.3 Practical skills

You may be asked to carry out a series of tests and record your findings. In this case the emphasis is on your ability to follow instructions carefully, to work cleanly and record complete and accurate observations.

Sometimes you will be expected to go further and to make inferences from your observations and then to use your knowledge of inorganic reactions to explain what you have noted.

Implementing

Carrying out the tests

Always follow the instructions carefully. Use small quantities of solids. Dilute solutions do not contain much of a chemical so if you add a lot of a solid you can easily hide the effect that you are looking for.

When using a teat pipette to add a solution, hold the tip of the pipette just above the open end of the test tube to avoid contaminating the reagent when you put the pipette back in the bottle.

When heating always start by warming gently while shaking to keep the contents of the test tube moving. Be very careful to avoid sudden boiling (bumping) which can eject chemicals from a test tube in a hazardous way.

Make sure that you can test for gases reliably either using a teat pipette or a delivery tube.

Recording observations

You should write down all that you observe: temperature changes, solids dissolving, precipitates forming, colour changes, vapours with a distinctive smell, gases bubbling off and the results of tests to identify the gases.

Take care not to confuse observations and inferences. Suppose you add some ammonia solution to a pale blue solution salt. What you see is that a

Only carry out practical work in the presence of a qualified supervisor.

Check safety before carrying out any practical procedure.

Hint

Always be careful to mix chemicals thoroughly.

Work cleanly to avoid false results.

milky blue precipitate forms at first, but on adding more ammonia solution the precipitate dissolves in the excess to give a deep blue solution. This is what you record as your observations. Under inferences you may then suggest an explanation for these observations in terms of what you know about the chemistry of copper(II) ions.

Analysing and drawing conclusions

Making inferences from your observations
Draw on your knowledge of inorganic reactions and on the tables of tests and results in this chapter to interpret your observations. Often you will find that individual tests are not conclusive. Take care not to read more into the results of a test than can be justified.

You will sometimes be asked to test a compound that contains ions that you have not worked with before, in which case you may not be able to explain all that you observe but are only able to draw general conclusions.

Explaining your observations
In addition to the tests described in this chapter you should be thoroughly familiar with the test tube reactions of the elements and compounds included in your course specification.

For the alkaline earth metals in group 2 you should know:

- the behaviour in water and dilute acids of the metals, metal oxides and metal carbonates
- the solubilities in water of the hydroxides, sulfates and nitrates
- the effects of heat on the carbonates and nitrates.

For the halogens in group 7 you should know:

- what happens on adding sodium hydroxide to aqueous halogens
- the effect of warming solid halides with concentrated sulfuric acid
- the displacement reactions when aqueous halogens are added to aqueous halide ions.

For the selected transition metals you should know:

- the colours in aqueous solution of the common oxidation states
- the effect of adding oxidising agents such as chlorine, potassium manganate(VII) or hydrogen peroxide to lower oxidation states
- the effect of adding reducing agents such as metals in acid, sulfur dioxide or iron(II) ions to higher oxidation states
- the colours of the common complex ions.

Hint

Record everything even if at the time you cannot explain what is happening.

If there is a colour change, always record the colour before and after the test.

Hint

Do not confuse 'colourless' (no colour) with 'clear' (transparent). A solution can be both coloured and clear.

If a solution turns 'white' during a test this means that a white suspension of a fine precipitate has formed.

Test Yourself

7.1 For each of the following tests identify the type of chemical reaction taking place, name the products and write a balanced equation for the reaction:

(a) testing for iodide ions with silver nitrate solution

(b) adding dilute hydrochloric acid to magnesium carbonate

continued ➢

Test Yourself *continued*

(c) using limewater to confirm the presence of carbon dioxide

(d) adding dilute sodium hydroxide solution to a solution of iron(II) ions

(e) testing for sulfur dioxide gas with paper moistened with an acidic solution of potassium dichromate(VI) ions

(f) adding ammonia solution to a precipitate of copper(II) hydroxide.

7.2 Identify the following salts and account for the observations:

(a) A white solid which colours a flame bright yellow. On heating the solid gives off a gas that turns acid dichromate(VI) solution green. No precipitate forms on mixing a solution of the salt with sodium hydroxide solution. A solution of potassium manganate(VII) turns from purple to colourless when added drop-by-drop to a solution of the salt.

(b) A white solid which sublimes on heating. A gas which turns universal indicator paper blue-green is evolved when the solid is warmed with sodium hydroxide solution. Adding dilute nitric acid followed by silver nitrate to a solution of the salt produces a white precipitate.

(c) A pale violet, double salt which, on mixing with calcium hydroxide and warming turns brown and gives off a pungent gas which turns moist red litmus paper blue. A solution of the salt reacts with sodium carbonate solution to give a gas which turns limewater cloudy white. After acidifying with nitric acid, the solution also gives a white precipitate with barium chloride solution. Adding ammonium thiocyanate to a solution of the double salt produces a very deep red colour.

(d) A white crystalline solid which colours a flame mauve. Warming the solid with concentrated sulfuric acid produces steamy fumes of a gas that turns blue litmus paper red mixed with an orange vapour. The solid is soluble in water. The solution does not change the colour of indicators. Mixing the solution with a solution of silver nitrate produces a cream precipitate which is insoluble in dilute ammonia but soluble in concentrated ammonia. The solution of the original solid turns orange on adding aqueous chlorine.

(e) A green solid which colours a flame green with flashes of blue. Heating the solid produces a vapour which condenses to a liquid that turns cobalt(II) chloride paper from blue to pink. The salt dissolves in water to give a blue solution which gives a white precipitate when acidified with dilute nitric acid and then mixed with silver nitrate solution. The green salt dissolves in concentrated hydrochloric acid to give a yellow solution that turns pale blue when diluted with water.

CHAPTER EIGHT

Organic reactions

8.1 Principles

You are not expected to have practical experience of all the organic reactions that you need to know about in theory. The following summaries concentrate on the more common reactions that you will meet in the laboratory. Check with your course specification to find out which compounds and tests you need to know for each main unit of your course.

Functional groups

A functional group is a group of atoms and bonds which gives a series of organic compounds its characteristic properties. The functional group in a molecule is responsible for most – though not all – of its reactions. This is because the hydrocarbon chain which makes up the rest of any organic molecule is generally inert to most common reagents, especially aqueous reagents.

Organic reactions used in tests

A reaction is only useful as a chemical test if it can produce an observable change.

Acid-base reactions
Organic acids are weak acids. Common compounds that show acidic behaviour are carboxylic acids and phenols. Phenols are much weaker acids than carboxylic acids.

Compounds with $-NH_2$ groups are bases. The order of base strength is: aliphatic amine > ammonia > phenylamine. These are all weak bases.

Acids and bases can be detected by the colour changes of indicators.

An organic acid that is insoluble in water is likely to be much more soluble in alkali when it reacts to form an ionic salt. Similarly an insoluble organic base is likely to be much more soluble in acid.

Acids which are as strong as, or stronger than, carboxylic acids produce carbon dioxide from carbonates. The mixture fizzes and gives off a colourless gas that turns limewater cloudy white.

Redox reactions
Two strong oxidising agents are acidic solutions of potassium dichromate(VI) and potassium manganate(VII). Potassium dichromate(VI) turns from orange to green when it reacts. Potassium manganate(VII) turns from a deep purple colour to colourless in acid solution.

Two other reagents that are less powerful oxidising agents are Fehling's solution and Tollen's reagent. These reagents are used to test for aldehydes.

Many of the test reagents mentioned in this chapter are hazardous. Before carrying out any tests, check that you are aware of the hazards and are taking suitable precautions to reduce the risks. Never work unsupervised.

Fehling's solution contains a deep blue copper(II) compound in alkaline solution. Aldehydes (and reducing sugars) reduce the reagent from copper(II) to copper(I) which appears as an orangey-red precipitate. At the start of the change the mixture first turns greenish because of the mixture of blue solution and orange precipitate.

Tollen's reagent is an alkaline solution containing colourless silver(I) ions as a complex with ammonia. Aldehydes reduce the silver(I) to metallic silver which can produce a silver mirror on clean glass or a very dark grey precipitate.

Addition reactions

The addition of bromine to C=C double bonds is the usual test for alkenes. The reaction produces a visible change because free bromine molecules are orange. The addition products with alkenes are colourless.

Hydrolysis reactions

A hydrolysis reaction involves the splitting of an organic molecule into two parts by reaction with water. Many hydrolysis reactions are slow but go faster on warming and in the presence of an acidic or basic catalyst.

Hydrolysis in the presence of alkali is used to test for halogenoalkanes and to study esters.

Addition–elimination reactions

One of the characteristics of carbonyl compounds (aldehydes and ketones) is a type of reaction in which addition is followed by elimination of water. The reactions of 2,4-dinitrophenylhydrazine with carbonyl compounds are addition–elimination reactions. They produces bright orange, solid precipitates. Measuring the melting of point of a 2,4-dinitrophenylhydrazine derivative can help to identify the aldehyde or ketone. For accurate results the derivative must first be separated and purified by recrystallisation (see pages 75–76).

Substitution reactions in compounds related to benzene

The most common laboratory example of a substitution reaction in the benzene ring is the reaction of bromine with phenol. This reaction is rapid at room temperature. The product is colourless, so the bromine is decolourised. The product is also insoluble in water so a white precipitate forms; this distinguishes the reaction from the addition of bromine to an alkene.

Complex-forming reactions

One of the characteristics of transition metal ions is that they form coloured complex ions. The formation of coloured complexes by iron(III) ions can be used to test for the simpler carboxylic acids and for phenols.

Hint

On shaking aqueous bromine with a liquid hydrocarbon the bromine molecules will largely move from the aqueous layer (which becomes paler) to the hydrocarbon layer (which turns orange) unless it is an alkene that reacts with the bromine so that it turns colourless.

8.2 Tests and observations

Preliminary tests

Preliminary observations and tests provide a general introduction to the characteristics of a compound. They can include:

Alkenes

Functional group: >C=C<

Physical properties	Solubility and acid–base character
Ethene, propene and the butenes are colourless gases. Common alkenes with more than four carbon atoms are liquids	Alkenes, like other hydrocarbons, do not mix with or dissolve in water. They have no acid–base properties

Functional group tests

Test	Observations	Inferences
Shake with dilute, aqueous bromine	Bromine rapidly decolourised. (Note that a hydrocarbon that is not an alkene will simply extract the bromine colour from the aqueous layer)	This test detects the presence of alkene double bonds but other compounds can decolourise bromine (see phenol)
Shake with a few drops of dilute, acidified potassium manganate(VII)	Purple colour quickly disappears	A compound that can reduce manganate(VII). Other organic compounds give this result

Halogenoalkanes

Functional group: chloroalkane   bromoalkane   iodoalkane

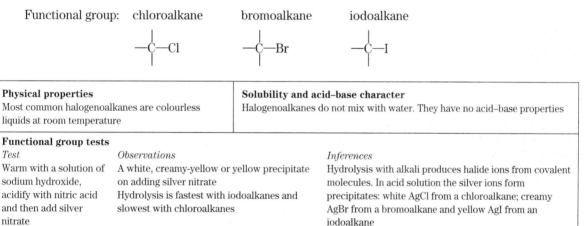

Physical properties	Solubility and acid–base character
Most common halogenoalkanes are colourless liquids at room temperature	Halogenoalkanes do not mix with water. They have no acid–base properties

Functional group tests

Test	Observations	Inferences
Warm with a solution of sodium hydroxide, acidify with nitric acid and then add silver nitrate	A white, creamy-yellow or yellow precipitate on adding silver nitrate. Hydrolysis is fastest with iodoalkanes and slowest with chloroalkanes	Hydrolysis with alkali produces halide ions from covalent molecules. In acid solution the silver ions form precipitates: white AgCl from a chloroalkane; creamy AgBr from a bromoalkane and yellow AgI from an iodoalkane

CHAPTER 8

Alcohols

Functional group: primary   secondary   tertiary

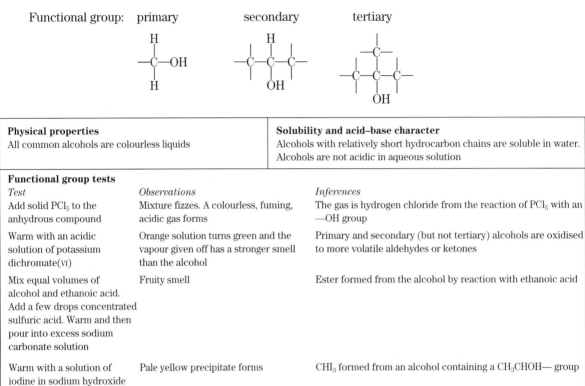

Physical properties	Solubility and acid–base character
All common alcohols are colourless liquids	Alcohols with relatively short hydrocarbon chains are soluble in water. Alcohols are not acidic in aqueous solution

Functional group tests

Test	Observations	Inferences
Add solid PCl_5 to the anhydrous compound	Mixture fizzes. A colourless, fuming, acidic gas forms	The gas is hydrogen chloride from the reaction of PCl_5 with an —OH group
Warm with an acidic solution of potassium dichromate(VI)	Orange solution turns green and the vapour given off has a stronger smell than the alcohol	Primary and secondary (but not tertiary) alcohols are oxidised to more volatile aldehydes or ketones
Mix equal volumes of alcohol and ethanoic acid. Add a few drops concentrated sulfuric acid. Warm and then pour into excess sodium carbonate solution	Fruity smell	Ester formed from the alcohol by reaction with ethanoic acid
Warm with a solution of iodine in sodium hydroxide	Pale yellow precipitate forms	CHI_3 formed from an alcohol containing a $CH_3CHOH—$ group

Aldehydes

Functional group:

Physical properties	Solubility and acid–base character
Methanal is a gas at room temperature. Ethanal boils at $21\,°C$. All other common aldehydes are colourless liquids	The simpler aldehydes such as methanal and ethanal are freely soluble in water. They have no acid or base properties in aqueous solution

Functional group tests

Test	Observations	Inferences
Warm with freshly prepared Fehling's solution	Mixture turns green, then the blue colour goes and an orange-red precipitate forms	Aldehydes reduce copper(II) ions in the reagent to copper(I) oxide
Warm with Tollen's reagent (ammoniacal silver(I) nitrate)	Silver mirror forms on clean glass	Aldehydes reduce silver(I) ions in the reagent to metallic silver
Add a solution of 2,4-dinitrophenylhydrazine	Thick yellow or red precipitate forms	Characteristic reaction of carbonyl compounds

Ketones

Functional group:

Physical properties	Solubility and acid–base character
All common ketones are colourless liquids at room temperature	The simpler ketones such as propanone mix freely with water but have no acid or base properties in aqueous solution

Functional group tests		
Test	*Observations*	*Inferences*
Warm with freshly prepared Fehling's solution	No reaction	Ketones do not reduce Fehling's solution
Warm with fresh Tollen's reagent Add a solution of 2,4-dinitrophenylhydrazine	No reaction	Ketones do not reduce Tollen's reagent
Add a solution of 2,4-dinitrophenylhydrazine	Thick yellow or red precipitate forms	Characteristic reaction of carbonyl compounds
Warm with a solution of iodine in sodium hydroxide	Pale yellow precipitate forms	CHI_3 formed from a ketone containing a CH_3CO— group

Carboxylic acids

Functional group:

Physical properties	Solubility and acid–base character
The simplest carboxylic acids, such as methanoic and ethanoic acid, are colourless liquids. Other common acids, such as ethanedioic acid and benzenecarboxylic acid, are solids	The simpler acids dissolve in water, they are weak acids. They give a solution with a pH below 7

Functional group tests		
Test	*Observations*	*Inferences*
Warm a little solid with dilute hydrochloric acid	Smell of vapour	Ethanoates give a strong smell of vinegar
Add a solution of sodium carbonate	The mixture fizzes and gives off a colourless gas that turns limewater cloudy white	Carbon dioxide given off by an acid
Neutralise and then add a neutral solution of iron(III) chloride	A red colour forms	Only methanoate and ethanoate ions give a reddish complex ion with iron(III) ions
Add ethanol and warm with a drop of concentrated sulfuric acid. Pour into sodium carbonate solution	Fruity smell detected	Formation of an ester by reaction of the acid with ethanol

Aromatic hydrocarbons (arenes)

Functional group:

Physical properties	Solubility and acid–base character	
Common hydrocarbons related to benzene are liquids	Benzene and related hydrocarbons do not mix with water. They have no acid–base properties	
Functional group test		
Test	*Observations*	*Inferences*
Ignite the hydrocarbon	Burns with a yellow and very smoky flame	Not a definitive test but helps distinguish benzene and related compounds from other hydrocarbons

Phenols

Functional group:

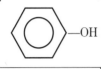

Physical properties	Solubility and acid–base character	
Phenols are solids at room temperature	Phenol itself is only slightly soluble in water but it dissolves in strong alkali to form soluble salts. It is a weak acid but not acid enough to react with carbonates	
Functional group test		
Test	*Observations*	*Inferences*
Add the phenol to a neutral solution of iron(III) chloride	Violet or purple colour	Formation of a purple complex ion indicates the presence of a phenolic —OH group

8.4 Practical skills

Test tube tests on organic reagents are sometimes used as planning exercises. Often, however, you are simply told what to do and the emphasis is on your ability to carry out the instructions in an appropriate way and then make relevant and complete observations. You may or may not be expected to explain what you observe.

The following sections suggest key points to attend to when you are being assessed on practical activities of this kind.

Planning

Working out a strategy for distinguishing chemicals

You might be asked to suggest a logical sequence of tests to identify a group of unknown compounds, given guidance about the nature of the chemicals in the group.

Presenting your plan

When you write out your plan should include:

- the series of tests in a logical sequence, possibly in the form of a flow chart

- the procedure with full details (including health and safety precautions bearing in mind that you will not know all the hazards if testing unknown compounds) for each test with a note of the expected observations if the test is positive

- an explanation of the chemistry underlying each test

- specific references to any sources of information you have been able to consult if not working under exam conditions (see Chapter 13).

Always show your plan to your teacher before starting any practical work.

Test Yourself

8.1 Outline a sequence of simple chemical tests that you could use to identify each of the compounds in this group of chemicals: propan-1-ol, propanone, propanoic acid, ethyl propanoate.

8.2 Suggest chemical tests to distinguish between these compounds:

(**a**) pent-1-ene, and pentane

(**b**) butan-1-ol and 2-methylbutan-2-ol

(**c**) propan-1-ol and propan-2-ol

(**d**) butanal and butanone

(**e**) propanone and pentan-3-one

(**f**) phenol and benzenecarboxylic acid (benzoic acid).

Hint

Note that it is sometimes important to include tests in your plans that will help eliminate functional groups.

Implementing

Following instructions

You may be asked to carry out a test that you have not practised before. If so carry out the test exactly as specified. When you have tried the test, check the instructions again to check that you have not missed out an essential step.

Only carry out practical work in the presence of a qualified supervisor.

You may have to use a simple test to check that you really have carried out a procedure as instructed. If, for example, you are told to acidify a solution you should mix well after addition of acid and then transfer a drop of the solution on to a small piece of blue litmus paper to check whether or not you have added enough acid.

If you are not sure whether a colour change is a change in the solution or due to the formation of a coloured precipitate, then you should separate some of the solid from the solution with the help of a centrifuge or by filtering.

If you are not sure whether or not an added reagent has caused a change, then repeat the test using pure water instead. The test with distilled water acts as a control. It is sometimes easier to look for differences.

Quantities

Always start by adding small quantities of chemicals. Never add more of a solid or solution than stated in the instructions.

If you find it hard to judge volumes, then use a measuring cylinder and marker pen to make a 1 cm^3 and a 5 cm^3 mark on one of the test tubes you are using.

If you add too much you may miss important clues. If a solid dissolves in excess acid you will fail to spot this if you add so much solid to the acid in a test tube that an excess of solid obscures an important observation.

Recording observations

Your notes should be concise, yet complete. Be on the look out for unusual changes and write down everything that you observe even if you do not understand all that is going on.

Be careful to distinguish your observations from their interpretation. Suppose you add some sodium carbonate to a solution of a unknown compound. What you see is that the mixture fizzes. The colourless gas turns limewater cloudy-white. These are the points that you record as your observations.

Analysing results and drawing conclusions

Making inferences from your observations

You have to use your knowledge of chemistry to decide what your observations mean. If a solution of an organic chemical reacts with sodium carbonate to produce a colourless gas that turns limewater milky, you can infer that the gas is carbon dioxide. From this you can conclude that the unknown compound is acidic and might be a carboxylic acid.

Drawing conclusions

You will not be able to draw definite conclusions from every test that you do. Sometimes you may not be asked for any inferences at all. Keep an open mind until you have carried out the complete series of tests. Bear in mind that you may be testing a chemical that you have never studied before. Your teacher, or the examiner, knows that this is the situation. The aim is to see whether or not you can make accurate observations and then suggest sensible interpretations based on your general chemical knowledge.

Do not get carried away and come to conclusions that you cannot justify from the evidence of your observations.

Using spectra to aid your interpretation

You are expected to be able to interpret simple mass spectra, infra-red spectra and nuclear magnetic resonance (nmr) spectra. You may be given spectra to analyse alongside the results of chemical tests. You will need to be able to refer to tables giving characteristic infra-red absorptions in organic molecules and giving chemical shifts for nmr spectra. Refer to your notes and your theory textbook for further details of how you do this.

Test Yourself

8.3 Identify the type, or types, of reaction taking place when there is a positive result with the following tests:

(a) testing for an alkene with bromine solution

(b) distinguishing ethanol and ethanoic acid with sodium carbonate solution

(c) testing for an aldehyde with Fehling's solution

(d) precipitating a derivative of a ketone with 2,4-dinitrophenylhydrazine

(e) testing for the presence of a phenol with neutral iron(III) chloride solution.

8.4 Account for the following observations as far as is possible from the evidence. You are not expected to identify the compound being tested:

(a) A colourless liquid does not mix with water. After warming a few drops of the liquid with aqueous sodium hydroxide, the resulting solution is acidified with nitric acid and then produces a cream-coloured precipitate on adding silver nitrate solution.

(b) A white solid chars on heating and gives off a vapour that condenses to a liquid that turns cobalt chloride paper from blue to pink. A solution of the solid turns universal indicator red. Mixing the solution with sodium carbonate solution gives a colourless gas that turns limewater cloudy. Warming a little of the solid with ethanol and a drop of concentrated sulfuric acid gives a product with a sweet smell detected on pouring the reaction mixture into cold water.

(c) A liquid burns with a yellow and very smoky flame. The liquid does not react with sodium carbonate solution but fizzes and gives white fumes with phosphorus pentachloride. The fumes turn blue litmus paper red.

CHAPTER NINE

Organic synthesis

9.1 Principles

Inorganic reactions between ionic compounds in solution are often fast. The reactions used to make organic compounds are generally slow because they involve bond breaking in covalent molecules. This means that the reactants in a synthesis have to be mixed and heated together for some time.

Stages in the preparation of an organic compound

Planning

Planning an organic preparation involves choosing a suitable reaction or sequence of reactions for making the product from available starting chemicals. Once this decision has been made it is possible to work out appropriate reacting quantities from the equation (see Chapter 2) and to decide on the conditions for the reaction. At the same time a risk analysis helps to ensure that there are no avoidable hazards and that the planned procedure minimises risks (see Chapter 11).

Carrying out the reaction

The reactants are measured out and mixed in a suitable apparatus. Cooling may be necessary while mixing some chemicals. Then it is often necessary to heat the reaction mixture for some time while preventing loss of chemicals with the help of a reflux condenser (see Figure 9.1).

When heating, choose the safest available method such as electric heating or heating with a water bath. Take particular care if heating with a flame.

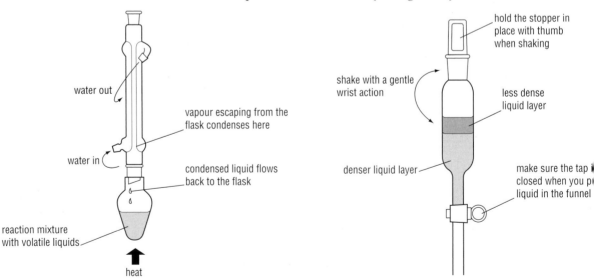

Figure 9.1 *Heating in a flask with a reflux condenser prevents vapours escaping while the reaction is happening. Vapours from the boiling reaction mixture condense and flow back (re-flux) into the flask*

Figure 9.2 *If the reactants do not all mix, but the reactio occurs at room temperature, then a possible purification procedure is to shake the reagents in a tap funnel*

Separating the product from the reaction mixture
Chemists sometimes talk about 'working up' the reaction mixture to describe the steps involved in separating the crude product. Distillation may be suitable if the main product is a liquid. If the main product is a solid it can be separated in an impure state by filtration.

Purifying the product
Products are generally contaminated with by-products of the main reaction and with some of the reagents used during the preparation.

A liquid product can be purified by shaking in a tap funnel with reagents that do not mix with the product but which can extract the impurities (Figure 9.2). Then drying and distillation gives a pure product.

Impure solids can be purified by recrystallisation (see page 90).

Checking the identity, purity and yield of the product
Measuring the melting point of a solid helps to check whether or not it is pure and to confirm its identity. Pure solids melt at a sharp temperature. Another useful technique for checking the purity of a solid is thin-layer chromatography. A pure compound produces just one spot on the developed chromatogram.

The final distillation of a liquid provides clues to the identity and purity of the product. During the distillation the main fraction should distil off over a narrow temperature range at the boiling point of the liquid.

Infra-red spectroscopy is a quick and convenient way to check the identity and purity of any organic product (solid or liquid). The spectrum of the product can be checked against a definitive spectrum in a database (see Chapter 13 for a website with spectra.)

9.2 Preparing a liquid product

The preparation of 1-bromobutane from butan-1-ol illustrates the main stages in the preparation and purification of a liquid product. The synthesis is based on a nucleophilic substitution reaction. The nucleophile is the bromide ion. With a primary alcohol the reaction goes on heating under acidic conditions.

Planning the preparation

Planning the preparation of an organic compound involves a lot of informed guesswork. Perfecting the conditions to get the best yield is often a matter of trial and error. Over the years this experience has been documented so that chemists, like cooks, can turn to recipe books to find out how to make common chemicals. Texts with preparative details indicate the likely yield and suggest suitable quantities of starting materials.

Chemists have found that it is possible to convert butan-1-ol to 1-bromobutane in one step using a mixture of sodium bromide and concentrated sulfuric acid. An excess of these two reagents ensures that as much of the alcohol reacts as possible.

Check safety before carrying out any practical procedure.

Planning includes a risk assessment (see Chapter 11) which is always necessary even when carrying out a tried and tested procedure.

Mixing the reagents and carrying out the reaction

Adding concentrated sulfuric acid to a mixture of butan-1-ol and sodium bromide is highly exothermic. So the mixture must be cooled while the acid runs in slowly from a tap funnel. Gentle swirling of the flask ensures that the reagents mix and cool (Figure 9.3).

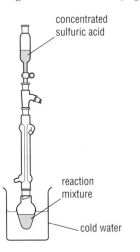

concentrated sulfuric acid

reaction mixture

cold water

Figure 9.3 *Cooling the reaction mixture while adding concentrated sulfuric acid*

Substitution of bromine atoms for the —OH in an alcohol involves breaking covalent bonds. This is a relatively slow reaction. The mixture is heated for about 45 minutes. A reflux condenser prevents the loss of volatile reactants and products (Figure 9.4).

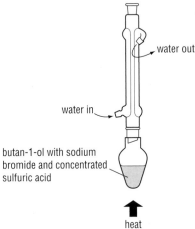

water out

water in

butan-1-ol with sodium bromide and concentrated sulfuric acid

heat

Figure 9.4 *Heating the reaction mixture in a flask fitted with a reflux condenser*

Separating the product from the reaction mixture

After cooling, the apparatus is rearranged for a simple distillation. Heating distils off the product together with acid fumes and some unchanged butan-1-ol (Figure 9.5).

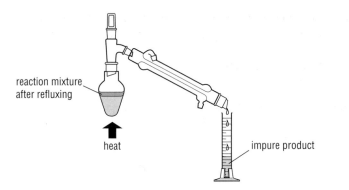

reaction mixture
after refluxing

heat

impure product

Figure 9.5 *Using
distillation to separate the
product from the reaction
mixture. Ionic salts are
involatile and do not distil.
Other molecules are volatile
and distil over with the
product*

Purifying and drying the product

The product does not mix with water. It can be purified by shaking with
aqueous reagents in a tap funnel. The denser 1-bromobutane can be run off
and retained each time while the aqueous reagent is thrown away.

Concentrated hydrochloric acid extracts unchanged butan-1-ol. Sodium
hydrogencarbonate then extracts and removes acids (Figure 9.6).

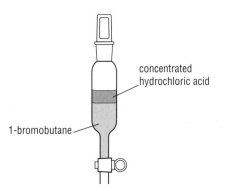

concentrated
hydrochloric acid

1-bromobutane

Figure 9.6 *Shaking first
with HCl(aq) and then with
$NaHCO_3$(aq) removes the
main impurities*

The product is run off from the tap funnel into a small flask. Adding a drying
agent such as anhydrous sodium sulfate or anhydrous calcium chloride
removes water (Figure 9.7). The product turns from a cloudy to a clear
liquid as water is removed.

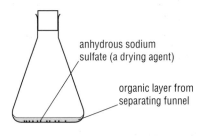

anhydrous sodium
sulfate (a drying agent)

organic layer from
separating funnel

Figure 9.7 *Drying the
product*

Final purification and identification

A final distillation gives 1-bromobutane free of other impurities. The boiling point of the product is 102 °C. Collecting the fraction which distils over between 100 °C and 104 °C gives a pure product so long as the distillation is carried out slowly (Figure 9.8).

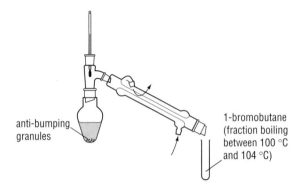

anti-bumping granules

1-bromobutane (fraction boiling between 100 °C and 104 °C)

Figure 9.8 *Final distillation to obtain the product free of by-products*

Measuring and accounting for the yield

Measuring the actual yield and comparing it to the theoretical yield is an important way of assessing any synthesis (see Section 2.4 on page 11). Chemists aim to make each reaction as efficient as possible. A perfectly efficient reaction would convert all the atoms in the starting materials into the desired products giving no by-products.

Most preparations give an actual yield that is less than the theoretical yield based on the balanced equation. There are several reasons why the overall yield may be low:

- The reaction may be incomplete: in this example, some unchanged butan-1-ol remains in this preparation even with an excess of sodium bromide and concentrated sulfuric acid.

- There may be side reactions which occur alongside the main reaction and use up some of the reagents to create by-products: in this preparation butan-1-ol reacts with concentrated sulfuric acid; some of the alcohol is dehydrated to an alkene and some turns into an ether.

- Some of the product is lost during the process as the chemicals are mixed, heated, distilled, transferred from one container to another, washed, dried and redistilled. Mechanical losses of this kind are unavoidable but can be limited by good technique.

9.3 Preparing a solid product

N-phenylethanamide (acetanilide) was once a popular analgesic marketed under the name antifebrin. Even though antifebrin is now considered too toxic for medicinal use, its discovery did stimulate the development of safer and more effective drugs. Paracetamol is a close relative of antifebrin.

Acylation of phenylamine produces antifebrin. The procedure illustrates techniques which can be used to make and purify organic solids.

It is safer to start with a salt of phenylamine because a solid salt is less volatile. Phenylammonium chloride, $C_6H_5NH_3^+Cl^-$, is a solid at room temperature. The preferred acylating agent is ethanoic anhydride which reacts with the free amine but not with the salt.

Mixing the reagents and carrying out the reaction

The first step is to dissolve a measured quantity of phenylammonium chloride in water and then add an excess of ethanoic anhydride (see page 12).

Adding a solution of sodium ethanoate converts the amine salt into phenylamine. The ethanoate ion acts as a base:

$$C_6H_5NH_3^+(aq) + CH_3COO^-(aq) \rightarrow C_6H_5NH_2(aq) + CH_3COOH(aq)$$

The phenylamine then reacts with ethanoic anhydride to produce the product which is insoluble in cold water and separates as a solid:

$$C_6H_5NH_2(aq) + (CH_3CO)_2O(aq) \rightarrow CH_3CONHC_6H_5(s) + CH_3COOH(aq)$$

Separating the product from the reaction mixture

Filtering under reduced pressure separates the impure product from the aqueous reagents (Figure 9.9).

⚠

Check safety before carrying out any practical procedure.

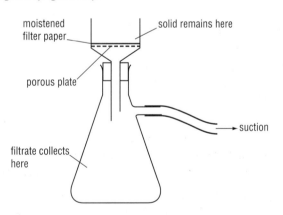

moistened filter paper

solid remains here

porous plate

suction

filtrate collects here

Figure 9.9 Separating the impure product by filtration

Purifying the product

The first stage of purification is to run a little cold water through the product on the filter paper. This washes away excess reagents.

Next the solid can be recrystallised from the minimum volume of hot water (Figure 9.10).

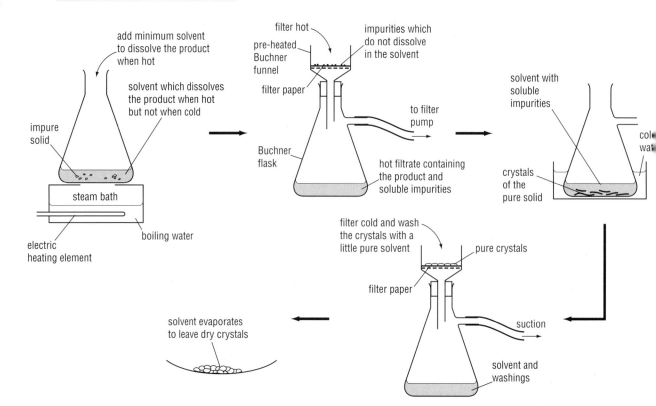

Figure 9.10 *Purifying a solid product by recrystallisation*

Final purification, identification and determination of the yield

The purified product is dried in air at room temperature. Drying can be quicker and more complete with the help of a desiccator. This is a sealed container with a drying agent in its base (Figure 9.11).

Figure 9.11 *Drying a solid in a desiccator*

Transferring the pure product to a weighed sample tube makes it possible to measure the actual yield.

Measuring the melting point of the dry solid helps to confirm the identity and purity of the product (Figure 9.12). Pure organic solids melt sharply at a temperature characteristic of the product.

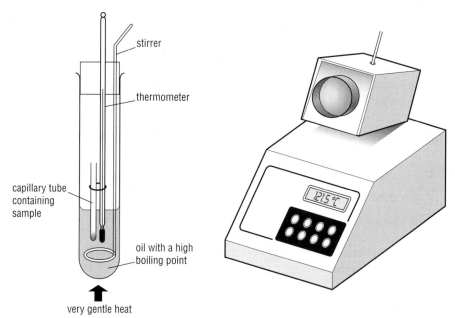

stirrer

thermometer

capillary tube containing sample

oil with a high boiling point

very gentle heat

Figure 9.12 *Two methods of measuring the melting point of a solid*

9.4 Practical skills

Planning

If asked to plan an organic synthesis you are likely to be given an outline of the process and then asked to work out the details.

Based on the summary given you should give full details of the preparation and purification of the product with sufficient detail which someone else, with similar experience to you, could follow successfully. You should use labelled diagrams to help to specify the apparatus required.

You should use the equations for the reactions to estimate suitable quantities of reactants where these are not given. You should also work out the theoretical yield of the product and, if given the likely percentage yield, also work out the mass of the product you would actually expect to prepare.

You should identify possible hazards and carry out a risk assessment (see Chapter 11).

If you have access to reference material it is important that you give full details of the sources you have used (see Chapter 13).

Always show your plan to your teacher before starting any practical work.

Implementing

You are expected to be proficient in all the practical procedures listed in your course specification. In part you will be assessed by the outcomes: the

Only carry out practical work in the presence of a qualified supervisor.

appearance of the product, the yield and the accuracy of its melting or boiling point. You may also be assessed by observation of your practical performance.

While carrying out the procedure you should keep a note of your main observations. Look out for colour changes and the appearance of precipitates.

Analysing and drawing conclusions

Generally the skills of analysis and drawing conclusions do not apply when carrying out a single organic synthesis. However, you might choose to make the same product by different routes using different reagents and conditions. If so, you would want to discuss the advantages and disadvantages of the methods you try, taking into account the ease of operation, the yield and the cost effectiveness of the alternatives.

Evaluation

Appearance of the product
Even the simple appearance of a product can give clues to the quality of the material. Pure organic solids are often white and crystalline. Pure liquids are clear and not cloudy.

Calculating yields
Measuring the percentage yield (see Section 2.4) allows you to assess how well you have carried out the preparation. Books of practical chemistry will often quote the expected percentage yield, with which you can compare your yield to decide how skilfully you have completed the synthesis.

Measuring melting and boiling points
Pure compounds have definite melting and boiling points. Values are tabulated for all common compounds.

Watching a solid melt gives clues to its purity. Pure solids melt sharply over a narrow range of temperatures. Impure solids soften and melt over a range of temperatures.

The method of mixed melting points is sometimes used to check an organic product. Mixing the product with a little of a pure sample of the same compound should not lead to any change in the observed melting point. If the product is not the expected compound, then the melting point of the mixture is very much lower than expected. Any impurity lowers the melting point markedly.

Fractional distillation of liquids does make it possible to estimate the boiling point of the main product, but this method of purification does not lead to such pure products as does the recrystallisation of solids. Therefore, crystalline derivatives are made that can be recrystallised and identified by their melting points (see Figure 8.1).

Chromatography
Thin-layer chromatography is a sensitive and quick way of detecting impurities in an organic product. A pure product moves the same distance up the chromatogram as a reference sample of the same compound. Also, developing a chromatogram of a pure product gives a single small spot on the chromatogram and not several spots.

Many organic compounds are colourless and so invisible at first on a thin-layer plate. The sample spots can be shown up either with the help of ultra-violet light or by exposing the plate to iodine vapour.

Spectroscopy

You are unlikely to have easy access to an infra-red spectrometer but, where available, the use of spectroscopy provides a very good method for analysing an organic compound. Reference spectra are available for all common compounds. By comparing the infra-red spectrum for your product with the spectrum for the compound in a database you can check on its identity and purity.

Test Yourself

9.1 Below is an account of the laboratory preparation of ethyl ethanoate. Study the account.

(a) Give an explanation of the purpose of each of the steps or chemicals added that are printed in italics.

(b) Look up the densities of the reactants and calculate the theoretical and percentage yields.

Mix 5 cm^3 ethanol and 5 cm^3 ethanoic acid in a flask. Add slowly, with *swirling and cooling*, 1 cm^3 of *concentrated sulfuric acid*. Fit a *reflux condenser* and boil gently *for 10 minutes*.

Rearrange the apparatus for distillation and *distil off everything boiling up to 82 °C*. Transfer the distillate to a separating funnel and *shake with sodium carbonate solution*. *Run off the aqueous layer* and next shake the product with a *concentrated solution of calcium chloride*.

Run off the aqueous layer again and then add a few small pieces of *anhydrous calcium chloride* to the product. When the liquid is *no longer cloudy*, decant it into a small distillation flask. Distil and collect the *fraction boiling between 74 and 79 °C*. Expect a yield of about 3 g.

CHAPTER TEN

Investigations

Hint

You should check to see whether or not your course specification requires you to carry out a complete investigation.

When picking an investigation check carefully with the assessment criteria to ensure that you will be able to score marks in all aspects.

Re-read the advice in Chapter 1 alongside the advice given in this chapter.

10.1 Choosing your investigation

You may be asked to choose your own topic for investigation or you may be provided with a problem brief. Either way you will have to decide on the specific questions that you are going to try to answer.

You are most likely to carry out an investigation as part of your A2 programme. This means that the questions you choose to study should relate to the theory in the A2 modules of your course, though you may well draw on AS knowledge in your planning and interpretation of results.

You will be wise to choose an investigation that grows out of some practical work that has arisen in your course. If you do so you will have practised the techniques; you will have a feeling for what is possible and some sense of a suitable procedure and scale of working.

Generally, it makes sense to pick an investigation that involves making measurements. This will help you to show off your skills of analysis and evaluation.

Aim to have a title for your investigation that is a question.

10.2 Planning

Once you have decided on your research question you should consider the possible ways of tackling the investigation. Consult various sources of information such as those listed in Chapter 13.

You must show your plans to your teacher before starting any practical work.

You will find accounts of practical procedures on websites. You should not simply copy these uncritically. Many are designed for university courses and will be impracticable in your laboratory. However, experimental instructions on web pages can give you an idea of suitable methods and quantities. If you decide to follow a procedure you find you must explain the experimental design and justify the method and quantities used.

You should consider trying out preliminary experiments on a small scale during the planning phase. This will give you a feel for the behaviour of the chemicals and help you to make decisions about quantities and concentrations.

If your investigation involves making measurements then you need to decide which ones are critical in determining the accuracy of your results. These are the measurements that need to be made as precisely as possible (see Chapter 12).

It is important that you carry out a risk assessment and decide how you will deal with any hazards (see Chapter 11).

10.3 Implementing

You will want to impress your teacher by the methodical and organised way that you carry out your planned procedures. Be careful to follow the safety precautions given in your risk assessment.

You are likely to modify some aspects of your planned procedure in the light of experience. This is expected and you will gain credit for explaining how and why you modified your approach in your report.

Only carry out practical work in the presence of a qualified supervisor.

10.4 Analysing evidence and drawing conclusions

Analyse your results with the help of relevant guidance in Chapters 2–9.

Show how you draw on what you have learned in a range of AS and A2 topics when analysing and interpreting your findings.

Checklist

In your analysis make sure that you:
* ★ state the chemical principles that are the basis for your analysis and write equations for all reactions
* ★ represent your data with appropriate charts, graphs or diagrams;
* ★ set out calculations in full, step-by-step
* ★ include units for all physical quantities
* ★ identify sources of measurement uncertainty and estimate the overall uncertainty in your final result (see Chapter 12)
* ★ quote your answer with significant figures that are consistent with the accuracy and precision of your work
* ★ draw conclusions from your results and comment on them in the light of chemical theory.

10.5 Evaluating evidence and procedures

Follow the advice given in Chapter 1 when evaluating your investigation. You should assess the reliability and precision of your data and evaluate the techniques used in your investigation.

Checklist

In your evaluation make sure that you:
* ★ comment on whether or not your results answer the question that you set out to investigate
* ★ discuss the reliability of your results in the light of your estimate of the overall measurement uncertainty
* ★ compare your results with predictions based on theory or results reported in published sources
* ★ review your original plan and suggests changes to the procedure that might give better results.

10.6 Writing the report

Your report should bring together all aspects of your investigations (see Sections 10.1–10.5).

Checklist

Your report should:
* ★ have a clear title preferably in the form of a question
* ★ include your plan for the investigation with an explanation for the choices you have made and calculations to justify any quantities
* ★ provide a risk assessment (see Chapter 11) and comment on how easy it was to implement the control measures
* ★ give a step-by-step account of any procedures illustrated with labelled diagrams

continued ➤

Checklist *continued*

★ give full records of all observations and measurements suitably tabulated or otherwise displayed

★ set out your analysis of the data collected, with graphs where necessary, plus all steps of calculations and the discussion relating findings to chemical theory; give your reasons if you choose to ignore some readings

★ show estimates of measurement uncertainty and other possible sources of experimental error (see Chapter 12)

★ state clearly your conclusions based on the evidence and relate these to the question being investigated

★ present your evaluation, commenting on the reliability of the findings, how your results compare with what might have been expected and how the procedure might be improved.

★ end with a list of references to secondary sources consulted (see Chapter 13).

CHAPTER ELEVEN

Hazards and risks

11.1 Health and safety

Hint

Remember this simple sequence when involved in practical work:

stop – think – do

Anything that can cause harm if things go wrong is a **hazard**. The chance (big or small) of harm actually being done is the **risk**. The risk depends both on the likelihood of something going wrong and on the seriousness of any possible injury.

You should minimise your exposure to risk. Always wear eye protection and whatever else is recommended. Follow instructions carefully. Take note of safety advice.

11.2 Chemical hazards and risks

There are several hazards associated with chemicals. Figure 11.1 shows the main hazards likely to arise in an advanced chemistry course.

Figure 11.1 The main hazards and their symbols

 Toxic chemicals may involve serious acute or chronic health risks and even cause death if breathed in, swallowed or absorbed through the skin

 Harmful chemicals may involve limited health risks if breathed in, swallowed or absorbed through the skin

 Corrosive chemicals destroy living tissues on contact

 Irritant chemicals are non-corrosive but may cause inflammation or lesions through immediate, prolonged or repeated contact with skin or eyes

 Highly or extremely flammable chemicals are liquids which are graded by the temperature at which they catch fire in air (flash point) and their volatility (boiling point) and solids which may catch fire after brief contact with a flame or which give off flammable gases in contact with water

 Oxidising chemicals which in contact with other substances produce a reaction that has a large heating effect and may ignite flammable substances

11.3 Risk assessment

Your teachers, and their employers, are responsible for carrying out risk assessments for most of your practical work. However you are expected to assess the risks when you are planning your own investigations.

Look carefully at your planned procedure and where you will be working. Decide:

- what chemicals are going to be used and in what quantities and concentrations (if in solution)

- how each hazardous chemical is to be used.

Identify any hazardous chemicals or hazards arising from equipment or procedures as well as the risks that might be involved in the work, such as:

- chemicals that can be splashed in the eyes, absorbed through the skin or breathed in

- chemicals that may spit or splash on heating, or catch fire

- reactions that are highly exothermic or may become violent

- hot apparatus which might cause burns

- glassware that leads to hazards if it blocks or cracks

- chemicals that need special measures for disposal at the end of a practical procedure

- faulty electrical equipment.

Evaluate the best way to minimise the risk of harm, through control measures. For example by:

- substituting a less hazardous chemical

- using a less hazardous form of the same chemical (for example, hydrated crystals instead of an anhydrous powder, or a more dilute solution)

- modifying a procedure to make it less risky, for example, by working in a ventilated fume cupboard to remove harmful vapours or cooling a reaction mixture that might otherwise be too violent

- heating liquids with a water bath or electric heating mantle instead of direct heating with a flame.

Always make sure that your teacher checks your plans and risk assessment before starting any practical work.

Hint

Refer to HAZCARDS when making your risk assessment (see page 105 for details).

Risk assessment form

Name of student:..

Title of practical activity:...

Summary of procedures:

Hazardous chemical being used or made, or hazardous procedure or equipment	Type of hazard and/or its consequences	Quantities and concentrations of chemical hazards	Precautions to control risks

Disposal of wastes

Signed by student: ....

Checked and signed by teacher: .... Date: ....

CHAPTER TWELVE

Measurement uncertainty

12.1 Types of measurement uncertainty

Errors of measurement are unavoidable differences between measured values and the true values. Often the true value is not known, so chemists have to assess the degree of uncertainty in their measurements.

There are **random errors** which cause repeat measurements to vary and to scatter around a mean value. Averaging a number of readings for the same measurement helps to take care of random errors. The smaller the random errors, the more precise the data.

There are **systematic errors** which affect all measurements in the same way, making them all lower or all higher than the true value. Systematic errors do not average out. Identifying and eliminating systematic errors is important for increasing the accuracy of data. Systematic errors can be corrected by calibration with another instrument that is known to be more reliable but may be too expensive for general use.

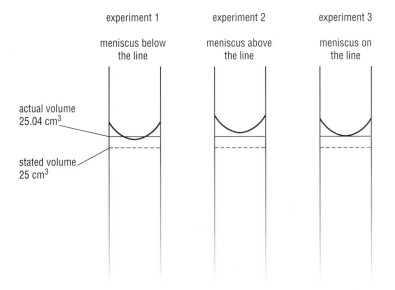

Figure 12.1 *Systematic and random errors in the use of a grade B pipette. The manufacturing tolerance for a 25 cm³ grade B pipette is ± 0.06 cm³. This can give rise to a systematic error. Every time an analyst uses the pipette the meniscus is aligned slightly differently with the graduation mark. This gives rise to random error*

Errors of measurement are not mistakes. Mistakes are failures by the operator and include such things as forgetting to fill the tip of a burette with the solution or taking readings from a sensitive balance in a draught. Mistakes of this kind should be avoided by the person doing the practical work. They should not be included in a discussion of the precision and accuracy of a set of results.

> **Hint**
>
> 'Measurement uncertainty' and 'experimental error' often mean the same thing. The term 'measurement uncertainty' is better. These are not mistakes but unavoidable variations in results.

CHAPTER 12

12.2 Sources of measurement uncertainty

Measurement uncertainty in chemical measurements can arise from a range of sources such as:

● the purity of the chemicals used as standards – suppliers of chemicals provide a range of grades of chemicals including analytical grades which are specially pure for quantitative work

● features of the measuring devices – such as the precision of balances or thermometers, the slight variations in the calibration of glassware, and the difficulty of reading from scales

● aspects of procedure – such as changes in the laboratory temperature which affect the volume of glassware which is calibrated at a particular temperature

● the behaviour of the person doing the titration – such as making the judgement of when the indicator colour change corresponds to the end-point.

Measuring volumes of solutions

Glassware

There are two common standards for glassware: grade A and grade B. The manufacturing tolerances allowed for grade B glassware are larger than for grade A glassware. Grade B glassware is cheaper and commonly used in schools and colleges for advanced chemistry courses.

There are three main sources of uncertainty when using volumetric glassware to measure the volumes of liquids:

● manufacturing tolerances mean that the graduation mark may be slightly different from the stated volume

● there are unavoidable variations in the judgement about the position of the meniscus relative to the graduation or scale marks

● glassware is calibrated at 20 °C and the volume will be slightly different is the working temperature if higher or lower than this.

Table 12.1 gives estimates of the sum of the systematic and random uncertainties associated with grade B glassware when used following standard procedures. For more details of how these estimates are made using statistical methods see *Introducing Measurement Uncertainty* from LGC Ltd.

Table 12.1 *Uncertainties associated with grade B glassware*

Glassware	Combined estimate of uncertainty for grade B glassware	Note
25 cm^3 pipette	± 0.05 cm^3	assuming that the pipette, like the other glassware has been used correctly
50 cm^3 burette	± 0.15 cm^3	allows for the uncertainties of two scale readings and in judging the end-point
250 cm^3 graduated flask	± 0.20 cm^3	never add a hot solution to calibrated glassware

Other measurements

Uncertainties in other measurements such as mass, temperature, time and pH depend on the quality and precision of the instruments available.

The overall uncertainty of a balance combines uncertainties arising from its precision, calibration and the readability of the scale. For a 2-place balance a reasonable estimate of the total uncertainty when used correctly in school or college is ± 0.01 g while for a 3-place balance the uncertainty can be taken as ± 0.001 g.

The uncertainty in measuring a temperature with a common 0–100 °C thermometer may be as much as ± 1 °C. This may not lead to significant error if the quantity used in calculating a result is the absolute temperature in kelvin. This would only introduce a percentage uncertainty of about $\pm 0.03\%$ in a value of 300 K. However, in a thermochemistry experiment to measure a temperature rise of 5 K the same uncertainty gives rise to a percentage error of 20% which is far too high. In thermochemistry it is generally essential to use a thermometer graduated in fifths or tenths of a degree. The measurement uncertainty with a thermometer of this kind can be taken to be ± 0.1 °C.

> **Hint**
>
> When planning an experiment, identify the uncertainties likely to make the larger percentage contribution to the overall uncertainty. Concentrate on keeping these as small as possible.

12.3 Calculating and combining uncertainties

In most quantitative experiments the final results are calculated from a number of measurements. The total uncertainty is determined by combining the individual uncertainties.

Uncertainties where measurements are added or subtracted

In an advanced chemistry course, when adding or subtracting measurements it is acceptable to add the measurement uncertainties.

Example

In a thermochemistry experiment the readings before and after a reaction were $19.3 \pm 0.1\,°C$ and $27.8 \pm 0.1\,°C$.

The temperature rise is $8.5\,°C$ and the total uncertainty is $\pm 0.2\,°C$.

This is likely to be an overestimate of the error in the difference because it is likely that both uncertainties are in the same direction: both too high or too low. This can be allowed for by combining the uncertainties a and b by calculating $\sqrt{a^2 + b^2}$.

Hint

Do not show the final result with more significant figures than can be justified in the light of the total uncertainty in the calculated value.

Uncertainties where measurements are multiplied or divided

In relationships where quantities are multiplied or divided, the values generally have different units. Here the first step is to calculate the percentage uncertainties:

$$\text{percentage uncertainty} = \frac{\text{uncertainty in the value}}{\text{value}} \times 100\%$$

The percentage uncertainties are ratios. They do not have units and can be added to arrive at a total percentage uncertainty for a calculated result.

Example

The table shows the results from a titration to determine the concentration of a solution of ethanoic acid:

Measurement	Value	Uncertainty	Percentage uncertainty
concentration of the standard solution of NaOH(aq)	$0.100\,\text{mol dm}^{-3}$	$\pm 0.001\,\text{mol dm}^{-3}$	$\dfrac{0.001}{0.100} \times 100\% = 1.0\%$
volume of ethanoic acid measured from a pipette	$25.00\,\text{cm}^3$	$\pm 0.05\,\text{cm}^3$	$\dfrac{0.05}{25.0} \times 100\% = 0.2\%$
volume of NaOH(aq) from a burette	$24.00\,\text{cm}^3$	$\pm 0.15\,\text{cm}^3$	$\dfrac{0.15}{24.0} \times 100\% = 0.6\%$

The concentration of ethanoic acid from these values $= 0.0960\,\text{mol dm}^{-3}$

Total percentage uncertainty $= 1\% + 0.2\% + 0.6\% = 1.8\%$

The total uncertainty in the calculated value $= \dfrac{1.8}{100} \times 0.096\,\text{mol dm}^{-3}$

$$\approx 0.002\,\text{mol dm}^{-3}$$

The concentration of ethanoic acid $= 0.096 \pm 0.002\,\text{mol dm}^{-3}$

CHAPTER THIRTEEN

References

13.1 Using reference sources

You are expected to consult more than one source of information when planning experiments and investigations. This can include books and periodicals, websites and CD-ROMs as well as experts other than your teacher or tutor.

It is very important that you keep a clear record of all the sources you consult at the time. See Section 13.2 to check on the information you need to note down about a source.

See Section 13.2 to check on the information you need to note down about a source.

Hint

Keep a careful note of the details of any sources you consult – at the time. This will save you trouble later on when you write a report.

Practical chemistry books for advanced courses

Barwick, V and Prichard, E (2003) *Introducing Measurement Uncertainty*, LGC Limited (ISBN 0948926198)

Hill, G and Holman, J (2001) *Chemistry in Context Laboratory Manual and Student Guide*, Nelson Thornes (ISBN 0174483074)

ILPAC (1997) *Advanced Practical Chemistry*, John Murray (ISBN 0719575079)

Nuffield Foundation (fourth edition, 2000) *Nuffield Advanced Chemistry Students' Book*, Pearson Education (ISBN 0582328357)

University of York Science Education Group (second edition, 2000) *Salters Advanced Chemistry Activities and Assessment Pack*, Heinemann (ISBN 0435631217)

Information about chemical and other practical hazards

ASE (third edition, 2001) *Topics in Safety*, Association for Science Education (ISBN 0863571042).

CLEAPSS® (2004) *Hazcards®*, CLEAPSS Science Publications

CLEAPSS® (2004) *Laboratory Handbook*, CLEAPSS Science Publications

CLEAPSS® (2004) *Student Safety Sheets*, CLEAPSS Science Publications

(The CLEAPSS publications are all available on the CLEAPSS Science Publications CD-ROM 2004 which is only accessible through schools and colleges which are members of CLEAPSS.)

SSERC (2002) *Haz Man CD2*, Scottish Schools Equipment Research Centre

Other useful reference books

Refer to your textbook for the theory related to your practical work.

Hunt, A (2003) *Complete A-Z Chemistry Handbook*, Hodder and Stoughton Educational (ISBN 0340872713)

Periodicals

The most useful periodical is *Chemistry Review* published by Philip Allan on subscription for advanced chemistry students. The magazine is published five times a year. The subscription is cheaper if bought through a school or college (see: www.philipallan.co.uk).

Websites

You can download the specification for your course from the website of the Awarding Body. You need to understand the assessment criteria for your practical work and for the content of examinations of the practical aspects of your course. These are the sites for the Awarding Bodies in England, Wales and Northern Ireland:

- AQA Chemistry – www.aqa.org.uk/qual/index.html

- CIEC Chemistry (Northern Ireland) – www.ccea.org.uk/gcesyll02.htm

- Edexcel Chemistry – www.edexcel.org.uk/qualifications/QualificationAward.aspx?id=48112

- Edexcel Nuffield Chemistry – www.edexcel.org.uk/qualifications/QualificationAward.aspx?id=48121

- OCR Chemistry – www.ocr.org.uk/OCR/WebSite/docroot/qualifications/qualificationhome/showQualification.do?qual_oid=2020&site=OCR&oid=2020&server=PRODUKTION

- OCR Salters Chemistry – www.ocr.org.uk/OCR/WebSite/docroot/qualifications/qualificationhome/showQualification.do?qual_oid=2021&site=OCR&oid=2021&server=PRODUKTION

- WJEC Chemistry (Wales) – www.wjec.co.uk/chemistry.html

You can find all the chemical data you need about elements and compounds including spectra at the *NIST Chemistry WebBook* – http://webbook.nist.gov/chemistry/. You can search the site by name or formula. If the chemical is normally a solid or liquid, search under 'Condensed phase'. If it is a gas search under 'Gas phase'. Learn to use the site by starting with a simple and familiar compound such as sodium chloride. Note that the site contains far more information than you need.

Another valuable source of chemical data is the *WebElements* website from Sheffield University – www.webelements.com/webelements/scholar/

The Nuffield Curriculum Centre runs a website, *Re:act* for advanced chemistry students – www.chemistry-react.org. The section on Practical Investigations and the hundreds of answers to student questions are helpful even if you are not taking the Nuffield course or using the Nuffield books. There are also many useful links to selected websites relevant to advanced chemistry courses.

Rod Beavons has an excellent website of particular value for those taking a course based on the Edexcel specification. It has good sections on inorganic reactions in qualitative analysis, and on organic preparations – www.rod.beavon.clara.net/chemistry_contents.htm

13.2 Adding references to your reports

Consulting reference material and seeking advice from other experts is not cheating. It is an essential part of planning any experiment or investigation. You must, however, acknowledge all the sources that you consult. Scientific articles and papers always end with a list of references. Your report must do so too. You get more marks if your references are clear and well chosen.

> **Hint**
>
> Gain marks by showing that you have made good use of appropriate sources of information and advice.

Books

State the author giving the last, or family, name followed by the initial. Then give the date of publication, the title of the book and the publisher. If you are word processing your report you should show the title in italics. In handwriting the title should be underlined. For example:

Barwick, V and Prichard, E (2003) *Introducing Measurement Uncertainty*, LGC Limited

Periodicals

Give the author and date with the title of the article you consulted, then give the details of the magazine with the volume number and edition plus the publisher. For example:

Carmody, M (2004) 'Well here it is! How can I purify it?' in *Chemistry Review*, Volume 13 number 4, Philip Allan

Websites

Give the title of the website, the web address and the date or dates on which you referred to the site. Give the web addresses of the parts of the site you consulted and not just the address of the home page. For example:

Re:act (Internet 12 December 2004) *Investigating the rate of the reaction of metals with acids*, www.chemistry-react.org/go/Tutorial/Tutorial_4425.html

Experts

Name any experts you have consulted other than your teacher or tutor. Give their names, their jobs and a sentence to explain how they helped.

ANSWERS

Chapter 2: Chemical amounts

2.1 (a) 0.1 g Mg gives 100 cm³ hydrogen under laboratory conditions.

(b) Yes, by cutting lengths from a cleaned strip of magnesium ribbon.

2.2 3.4 g of magnesium carbonate

2.3 About 12 cm³ of the alkene and 6 cm³ bromine

2.4 x = 6

2.5 (a) According to the equation 1 mol (7.0 g) Li gives 0.5 mol $H_2(g)$ which has a volume 12 000 cm³ under laboratory conditions. So in theory 0.021 g should give 36 cm³ gas.

(b) The metal surface may already be slightly oxidised while being stored under oil. Not all the oil may have been removed by the treatment with hexane. Both would tend to reduce the mass of metal available to react with water.

2.6 Theoretical yield = 9.6 g
Percentage yield = 65%

Chapter 3: Volumetric analysis

3.1 Loss in mass is 12.1%

3.2 $H_3PO_3(aq) + 2NaOH(aq) \rightarrow$
$Na_2HPO_3(aq) + 2H_2O(l)$

3.3 n = 2

3.4 Concentration = 0.01926yh mol dm^{-3}

3.5 Iron(II) concentration = 0.0720 mol dm^{-3}
Iron(III) concentration = 0.0180 mol dm^{-3}

3.6 Copper(II) is reduced to copper(I)

3.7 Hardness expressed as grams calcium carbonate per litre = 0.292 g dm^{-3}

Chapter 4: Enthalpy changes

4.1 Plan to use similar apparatus and quantities to that illustrated elsewhere in the chapter. Note that both magnesium metal and magnesium oxide react with hydrochloric acid to form a solution of magnesium chloride. This allows you to draw up a Hess's law cycle. Calculate the masses of metal and oxide that will react to give the same amount of magnesium chloride in each case. Make sure that in both reactions the acid is comfortably in excess. Note too that for the calculation you need to look up the value for the enthalpy of formation of water.

4.2 $\Delta H_{neutralisation} = -56.4 \text{ kJ mol}^{-1}$

4.3 $\Delta H_{solution} = +17 \text{ kJ mol}^{-1}$

Chapter 5: Reaction kinetics

5.1 Values of $\frac{1}{t}$ in s^{-1}: 0.017, 0.033, 0.050, 0.067, 0.083.
Studies of this reaction suggest that the rate is proportional to the concentration when the acid concentration is low; but that the rate is proportional to the square of the concentration when the acid is more concentrated. So the order depends on the concentration of the acid.

5.2 (a) $E_a = 35 \text{ kJ mol}^{-1}$

(b) The presence of a catalyst leads to a reaction pathway with a lower activation energy.

(c) One possibility is that when the catalyst is present the iron(III) oxidises iodide ions to iodine and that the peroxodisulfate(VI) then oxidises the resulting iron(II) back to iron(III) thus regenerating the catalyst. Perhaps the negative peroxodisulfate(VI) ions react more readily with positive iron(II) ions than with negative iodide ions.

Chapter 6: Equilibria

6.1 (a) The range is from:
pH = 4.1 + log 1/9 = 3.1
to:
pH = 4.1 + log 9/1 = 5.0

(b) Prepare a buffer solution by mixing 5.0 cm³ of 0.020 mol dm^{-3} methanoic acid and 5.0 cm³ of 0.020 mol dm^{-3} sodium methanoate. Add 10 drops bromophenol blue indicator. Use the set of tubes in figure 6.2 to find out the pH of the buffer solution. Since [HCOOH(aq)] = [HCOO$^-$(aq)], the pH of the buffer solution = pK_a of methanoic acid.

6.2 $K_c = 0.204$

6.3 $pK_a = 4.2$.

   $K_a = 6.3 \times 10^{-5}\,\text{mol dm}^{-3}$

Chapter 7: Inorganic reactions

7.1 (a) Ionic precipitation to form silver iodide.

   $Ag^+(aq) + I^-(aq) \rightarrow AgI(s)$

   (b) Acid–base reaction forming magnesium chloride.

   $MgCO_3(s) + 2HCl(aq) \rightarrow$
   $\qquad MgCl_2(aq) + CO_2(g) + H_2O(l)$

   (c) Acid–base and ionic precipitation forming calcium carbonate. Overall this can be summarised as:

   $Ca^{2+}(aq) + 2OH^-(aq) + CO_2(g) \rightarrow$
   $\qquad CaCO_3(s) + H_2O(l)$

   (d) Ionic precipitation producing iron(II) hydroxide.

   $Fe^{2+}(aq) + 2OH^-(aq) \rightarrow Fe(OH)_2(s)$

   (e) Redox reducing dichromate(VI) to chromium(III). Combine these two half-equations for the overall balanced equation.

   $Cr_2O_7^{2-}(aq) + 14H^+(aq) + 6e^- \rightarrow$
   $\qquad 2Cr^{3+}(aq) + 7H_2O(l)$
   $SO_2(aq) + 2H_2O(l) \rightarrow$
   $\qquad SO_4^{2-}(aq) + 4H^+(aq) + 2e^-$

   (f) Ionic precipitation to form a precipitate of copper(II) hydroxide followed by the formation of a stable complex which means that the precipitate redissolves to give a deep blue solution.

   $Cu^{2+}(aq) + 2OH^-(aq) \rightarrow Cu(OH)_2(s)$
   $Cu(OH)_2(s) + 4NH_3(aq) \rightarrow$
   $\qquad Cu(NH_3)_4^{2+}(s) + 2OH^-(aq)$

7.2 (a) Sodium sulfite, Na_2SO_3

   (b) Ammonium chloride, NH_4Cl

   (c) Ammonium iron(III) sulfate, $NH_4Fe(SO_4)_2$

   (d) Potassium bromide, KBr

   (e) Hydrated copper(II) chloride, $CuCl_2$

Chapter 8: Organic reactions

8.1 Mix with water: propanoic acid mixes with water to give an acidic solution; propan-1-ol and propanone mix with water but the solution is neutral; ethyl propanoate does not mix with water.

Add PCl_5 to the pure liquids: only the two –OH compounds (propan-1-ol and propanoic acid) react and give acid fumes.

Add a solution of 2,4-dinitrophenylhydrazine: only propanone gives an orange precipitate. Warm with sodium hydroxide solution the liquid that does not mix with water: the oily liquid slowly breaks down and mixes with the water (as ethyl propanoate hydrolyses to ethanol and propanoic acid).

8.2 (a) Shake with a solution of bromine. The alkene decolourises the bromine.

   (b) Warm with acidic sodium dichromate(VI). With the primary alcohol the oxidising agent turns from orange to green.

   (c) Warm with acidic sodium dichromate(VI). Distill off the product of oxidation. The distillate from propan-1-ol gives a positive reaction with Fehling's solution.

   (d) Warm with Fehling's solution. The reagent loses its blue colour and gives a brick-red precipitate with the aldehyde but not the ketone.

   (e) Warm with a solution of iodine in sodium hydroxide. Yellow precipitate of CHI_3 from propanone.

   (f) Make a solution in warm water. Adding a solution of neutral iron(III) chloride gives a violet colour with phenol. Adding a solution of sodium carbonate produces bubbles of gas with benzenecarboxylic acid.

8.3 (a) An electrophilic addition reaction.

   (b) An acid–base reaction in which a carboxylic acid protonates carbonate ions forming carbonic acid, which decomposes to carbon dioxide and water.

   (c) A redox reaction in which copper(II) is reduced to copper(I).

   (d) An addition–elimination reaction.

   (e) A complex forming reaction.

8.4 (a) The silver bromide precipitate suggests that bromide ions have been produced by hydrolysis of a covalent bromide. Bromoalkanes are immiscible in water. Nucleophilic substitution by reaction with aqueous hydroxide ions produces an alcohol and bromide ions.

(b) Carboxylic acids form sweet smelling esters on reaction with alcohols in the presence of an acid catalyst. Carboxylic acids give off carbon dioxide on reaction with aqueous carbonate ions. Some solid organic acids decompose to give carbon on heating. They may also be hydrated and give off water vapour on heating.

(c) Compounds related to benzene burn with a smoky flame. Compounds with –OH groups react with phosphorus pentachloride giving off hydrogen chloride gas, which is acidic. This might be a compound with a benzene ring and an alcohol –OH group attached to a side chain.

Chapter 9: Organic synthesis

9.1 (a) *swirling and cooling* – adding the concentrated acid to the other reagents is highly exothermic; mixing and cooling avoids excess local heating

concentrated sulfuric acid – this is the catalyst for the reaction
reflux condenser – this prevents vapours escaping during heating
for 10 minutes – the reaction between the covalent molecules is relatively slow even on heating
distil off everything boiling up to 82 °C – this separates volatile organic compounds from less volatile inorganic compounds
shake with sodium carbonate solution – this neutralises acids and removes them from the organic layer, including unchanged ethanoic acid
Run off the aqueous layer – this separates the impure product from the salts of acids in water
concentrated solution of calcium chloride – this absorbs any unchanged ethanol
anhydrous calcium chloride – this is a drying agent
no longer cloudy – organic liquids are cloudy when moist but clear when dry
fraction boiling between 74 and 79 °C – this is the fraction which contains the product which boils at 77 °C

(b) Ethanol is the limiting reagent (just). Theoretical yield = 7.5 g ester. Actual yield = 40%

INDEX